WIDENING THE FOOD BASKET

About the Authors

Mahendra K. Satapathy, is presently working as Professor in Science Education(Botany) and Dean, Academic at Regional Institute of Education (NCERT), Bhubaneswar. After passing B.Sc.(Hons), M.Sc. (Botany) from Utkal University, Bhubaneswar with first rank and University gold medal, Prof. Satapathy completed his Ph.D from National Rice Research Institute (NRRI), Cuttack. Having worked as a Senior Post-Doctoral Fellow at International Rice Research Institute (IRRI), Manila, Philippines, British Council Fellow at Wye College (University of London) and a Fulbright Fellow in University of Minnesota, USA, he has been working as Professor in National Council for Educational Research and Training (NCERT), New Delhi for more than last thirteen years. With a Teaching and Research Career of 33 years in Post Graduate Department of Life Science, Dr. Satapathy has headed the Departments of Science and Mathematics and Department of Extension Education, besides being felicitated as the Best teacher in Botany in Odisha. Prof. Satapathy has published a large number of Research Papers, and authored seven books and supervised a couple of students for Ph.D in Botany, Environmental Science and Education. His fields of interest include Science Education, Environmental Education and Sustainability besides Educational Innovations in the country.

Sidhanta Sekhar Bisoi, is presently working as Senior Project Fellow in the Department of Botany, Regional Institute of Education (NCERT), Bhubaneswar, Having completed his M.Sc. and M.Phil. in the field of Biodiversity and Conservation of Natural Resources from Central University of Orissa, Koraput, and M.Sc. in Botany from Berhampur University, Odisha, he has published a series of research papers in International journals of repute, besides being felicitated as a recognized reviewer for a couple of reputed national/international journals approved by UGC. He has presented his research work(s) in a number of national and international conferences/seminars/workshops. His areas of interest include Ethno-botany, Plant Diversity Studies, and Biodiversity Conservation. Currently, Mr. Bisoi is working in the area of Ethno-botany with reference to Indigenous Knowledge and Natural Resource Conservation.

WIDENING THE FOOD BASKET

Underutilized and Underexploited Food Resources

Mahendra K. Satapathy
Sidhanta S. Bisoi

NEW INDIA PUBLISHING AGENCY
New Delhi – 110 034

NEW INDIA PUBLISHING AGENCY

101, Vikas Surya Plaza, CU Block, LSC Market
Pitam Pura, New Delhi – 110 034, India
Email: info@nipabooks.com
Web: www.nipabooks.com

For customer assistance, please contact
Phone: + 91-11-27 34 17 17 Fax: + 91-11- 27 34 16 16
E-Mail: feedbacks@nipabooks.com

ISBN : 978-93-89130-22-5

Composed and Designed by NIPA.

Digitally printed all across the globe.

Dr. S.K. Dash

President, UAS Lab. & Adjunct Professor
Department of Food Sciences & Nutrition
South Dakota State University
USA

Foreword

Food energy intake depends on many factors, most importantly on availability of food of appropriate quantity and quality at affordable prices, besides purchasing power of the people. Though over the last two decades, agricultural productivity has stagnated internationally, the consumption of agricultural products has steadily increased with fast rise in population and increase in income level.

Though at national level in India, we have solved the problem of food security which is reflected in mounting buffer stocks, yet there are millions of food insecure and under nourished people in the country. There is an extremely high prevalence of hunger and malnutrition. As we are celebrating our 71st years of Independence, the rampant malnutrition, anaemic mothers and stunted children indicate our failure to meet the challenges of hidden hunger caused by deficiency of micro nutrients such as iron, iodine, zinc, vitamin A and vitamin B_{12} affecting about one fourth of our total population. The International Food Policy Research Institute, Washington has placed India in the 103th position out of 118 countries in Global Hunger Index, with India standing below all other south Asian nations, except Pakistan. Very recently, FAO, the UN food agency has warned that protecting the world's food supply is critical as the Biodiversity that underpins our food systems, is declining at all levels around the world pushing the planet's population towards chronic hunger

Though there are various means of meetings the challenges of food insecurity and malnutrition, through plant breeding and other biotechnological processes such as Tissue culture, distant hybridization etc., one of the earliest ways is to search for traditional plant resources growing in the wild and are rich in various nutrients. The authors are to be congratulated that through wide survey and research they have identified more that 100 plant species that are rich in proteins, amino acids, various micronutrients and vitamins and have reported in the present book "Widening the Food Basket: Underutilized and Underexploited Food Resources". Besides detail descriptions about each plant species and its ecology, they have spelt out the nutritional value, along with the picture of the plant with its local as well as botanical name.

I am sure the book will be popular among Nutritionists, Biologists, Agricultural Scientists, besides students and teachers of Botany, Agriculture and life Sciences.

-Sd-
S.K. Dash

Preface

Food is the basic need of every individual and stands first among the hierarchical needs of human being. As such, it is the right of every person to have regular access to sufficient nutritionally adequate and culturally acceptable food for an active healthy life.

Following the Green revolution that commenced from mid 1960's, India's per capita food availability went up considerably. However with decline in area under cultivation, because of urbanisation and industrialization, the production failed to keep up with population growth and consequently per capita availability declined from 500 gms per capita per day to less than 400 gms reaching close to 1950's.

The United Nation's Food and Agriculture Organisation (FAO) has observed that almost one- sixth of all humanity is suffering from hunger. According to a recent report from UNO, of about 900 million people who are under extreme poverty, 300 millions are from India that stands 103rd position in the global hunger index. At certain places and remote areas, chronic food insecurity persists and families often go empty stomach and eat on alternate days.

Through the National Food Security bill enshrines freedom from hunger and malnutrition as a fundamental right, an important issue related to food security is that poor people need to have enough purchasing power to have an access to sufficient and nutritious food grains. Mere availability of food grains does not ensure food security. Thus being grilled under poverty, people often search for wild plants and their parts such as fruits, seeds, roots, leaves, tubers etc. to meet their hunger.

Besides hunger, India has the largest number of malnourished people in the world amounting to about one third of the total two billion people worldwide. Malnutrition in terms of deficiency of essential amino acids (lysine, methionine, folic acids), vitamins (vitamin A and D) and microelements (such as iron, zinc, magnesium etc.) known as hidden hunger affects the developmental process including cognition level, blindness and immunological malfunctioning etc. Most of the malnourished individuals include infants, children and women in resource poor families. About 200 children per day die in India because of malnutrition.

The possible cause could be decline of food basket diversity especially of traditional grains, fruits, vegetable, berries etc. The Sustainable Development Goal-2 that aims to end hunger, achieve food security and promote sustainable agriculture" is a priority area for India. Several strategies (multisectoral approaches) are being pursued to meet the dietary needs including malnutrition through distant breeding process, dietary diversification, pharmaceutical preparations, bio-fortification etc. However, one of the easiest ways is to search for potential sources in the nature. Interestingly many plants growing in the wild are rich in nutrients and are traditionally in use by tribal and rural people.

In the present book, from our wide survey, we have gathered a series of such wild plants growing in diverse environments and having richness in nutrients in one form or the other, protein, amino acids, vitamins or micro nutrients. Each plant species has been spelt out in detail with its local and scientific names, ecology, botanical description, type of part in use, nature of nutrient(s) present and preparations made from it. Some non-traditional preparations of domesticated plants has also been described keeping their potential nutritional value.

We wish that book will be useful to a variety of readers such as students and teachers of Botany, Agriculture, Life Sciences, Environmental Sciences, Nutritionists, general public besides spreading awareness about availability of cheap nutritious food resources among key target groups including women, children and tribals.

It might have so happened that certain plants species having good nutritional value might have been missed inadvertently from our list. We request all the readers that in case any new plant species is noted, to kindly get it to our notice so that we could incorporate the species in the next edition of the book.

We acknowledge with thanks in Principal, Regional Institute of Education, Bhubaneswar and the Director, NCERT, New Delhi for their support and encouragement of the present peace of work.

Authors

Contents

Introduction

Food is the first among the hierarchical needs of a human being. A sufficient balanced diet is the most important contribution to human health and well being. All countries in the world have been striving hard to have self sufficiency in food grain production.

Following green revolution that commenced from mid sixties, India not only became a net exporter of food grain, but per capita food availability went up considerably from 390 gms per capita per day to a peak of 510gm in 1991. India with only 3% of the world's total land feeds 17% of the world population. With the spectacular success, India could become a role model for many developing nations in improving agricultural productivity. However with decline in area under cultivation because of urbanization and industrialization, the production in India has failed to keep up with population growth and consequently per capita availability of food grain has come down to less than 400 gms per day over the last two decades.

At International level, agricultural productivity has stagnated while the consumption of agricultural products has steadily increased with increase income levels and population growth. With the planet struggling to provide us with enough resources to sustain its 7 billion people, one in every seven people in the world go to bed hungry and more than 20,000 children under the age of 5 die daily because of hunger.

In most countries of the world, malnutrition is seen in the form of under nutrition that is caused by a diet lacking in adequate calories and protein. India is having the largest number of malnourished people in the world amounting for one third of the two billion people worldwide. Micronutrient malnutrition especially, iron, zinc, magnesium and manganese deficiency distress occurs over fifty percent of the world population and it is further increasing in developing countries. More than 5 million childhood death occur due to micronutrient malnutrition every year.

Though India is one of the largest producers of the food grains in the world, it has nearly 300 million people who sleep without two square meals a day. At certain places and in remote areas chronic food insecurity persists and families do eat on alternate days. Being grilled under poverty, they often search for wild plants and their parts such as fruits, seed, leaf, roots, tubers etc. to meet their hunger.Besides hunger, India has the largest number of malnourshied people in the world amounting for one third of the two billion people worldwide. Malnutrition includes deficiency of essential amino acids such as lysine, methionine, folic acids and micro elements such as iron, zinc, iodine and vitamins that affect developmental process including cognition besides blindness and immunological malfunctions. Since these symptoms are not associated with inadequate diet, it is known as hidden hunger. Most of the malnourished individuals include infants, children and women in resource poor families. About half of all death in children under the age of five years is because of malnutrition. About 200 children die in India per day because of malnutrition.

Several strategies such as breeding process, dietary diversification, pharmaceutical preparations, bio-fortification etc. are being pursued to meet the dietary need including malnutrition. However, one of the easiest ways is to search for the potential sources in the nature. Interestingly many plants rich in nutrients do grow in the wild and are traditionally used by rural and tribal people.

The present book carries a list of about 112 of such wild plants that are rich in nutrients in one form or the other, protein, amino acids, vitamins and micronutrients. Each plant species along with its picture has been described separately in detail with their local and scientific name, ecology, botanical structure, type of the plant part, nature of nutrients present and preparation made out of it in the following pages in the text for the interest of the readers.

1

Achyranthes aspera L.

Hindi Name: Chirchita

English Name: Prickly Chaff Flower

Family : Amaranthaceae

Flowering : October-December

Fruiting : December-January

Botanical description

The plant is an erect, straggling or subscandant herb. The leaves are ovate, elliptic, obtuse, acute and finely attached to the stem on both sides. The flowers are greenish white in colour, bracteolate having membranous blade and seeds are reddish brown in colour, sub-cylindric at apex and rounded at the base.

Ecology

It is a fairly common weed distributed throughout India and growing on waste lands.

Use: Young leaves and shoots are collected, roasted and then eaten

Nutritional Value: The plant is rich in vitamin B complex and protein. The nutritional value of the plant is described below. 100g of the plant contains approximately:

Carbohydrate 8.6g, Protein 10.2g, Fat 2.9g

Vitamin B_{1-}0.27mg, B_2-0.28mg, B_3-0.58mg, B_6-0.27mg

Mineral Ca-661mg, Fe-21.3mg, P-17.6mg

Economic value: The plant has also got medicinal value and is used against vomiting and dysentery.

Achyranthes aspera L.
Vernacular names
- Apamaranga (O)
- Prickly chaff flower (E)
- Chirchita (H)

2

Acalypha indica L.

Hindi Name: Kuppi

English Name: Indian Copper Leaf

Family : Euphorbiaceae

Flowering: July-September

Fruiting: September-December

Botanical description: The plant is a small erect herb, growing up to 60 cm or more. The ascending branches are angled and velvet-hairy. Leaves are broadly ovate, nearly triangular, rather coarsely toothed. Leaf stalks are as long as or longer than the 3-5 cm long blades. Flowers are stalkless, borne on erect axillary spikes longer than the leaves. Male flowers are minute, crowded distally. Female flowers are scattered along the inflorescence axis, each subtended by a conspicuous semicupular leaf-like toothed green bract nearly 7 mm long. Capsule is bristly and1 mm broad.

Ecology: It is mostly found in hot, low to medium altitude, sandy margins of rivers, seasonal water courses and pans, usually in shade of thickets, on rocky hillsides and often in disturbed ground and as a weed of cultivation.

Use: Leaves are fried with oil, and eaten

Nutritional Value: The plant is rich in vitamin and minerals. The nutritional value of the plant is described below. 100g of the plant contains about:

Carbohydrate 5.2g

Protein 8.2g

Vitamin B_1-0.38mg, B_2-0.18mg, B_6-0.97mg

Mineral Ca-889mg, Fe-74.3mg, P-12.6mg

Economic Value: The leaves are widely used in traditional medicines against skin parasites and other skin problems

Acalypha indica L.
Vernacular names:
- Mukta jhuri (O)
- Indian Copperleaf (E)
- Kuppi (H)

3

Aerva lantana (L.) Juss

Hindi Name: Chaya

English Name: Mountain Knot Grass

Family : Amaranthaceae

Flowering: August-October

Fruiting: November-January

Botanical description: The plant is a perennial sub-erect, diffuse or prostrate herb, growing upto 0.3-2 meter high, branched from the base and often also from above. Stem and branches are densely woolly with whitish or yellowish, shaggy hairs. Leaves are alternately arranged, nearly circular to lanceo shaped-elliptic, wedge-shaped at the base, rounded to sharp at the tip. Flower spikes are stalkless, solitary or usually in clusters in leaf axils, silky white to creamy in colour, forming a long inflorescence leafy to the ultimate spikes.

Ecology: This plant is commonly found in open forests on mountain slopes, on waste and disturbed ground and coastal scrubs. The plant is widespread in drier parts of the tropics and subtropics.

Use: Leaves are collected, fried and then eaten

Economic value: The plant has got medicinal importance and used against swellings and eye diseases.

Aerva lantana (L.) Juss
Vernacular names:
- Paunsia (O)
- Mountain Knot Grass (E)
- Chaya (H)

4

Allium cepa L.

Hindi Name: Pyaz(H)

English Name: Onion (E)

Family : Amaryllidaceae

Flowering: December-January

Fruiting: January-February

Botanical description: It is a herb. The leaves of the plants are cylindric, fistular and look like tubes. Flower of the plant is white or greenish white with a short pedicel.

Ecology: The plant is commonly cultivated along the river banks during winter season. The plant loves to grow in sandy and loamy soil.

Use: Young leaves and shoots are collected, fried then eaten. The leaves and the flowering stalks are used for various preparations and curries.

Nutritional Value

The plant is rich in vitamin and minerals. The nutritional value of the plant is described below. 100g of the plant contains:

Carbohydrate 9.4g

Protein 2.25

Fat 6.3g

Vitamin 18.75g

Mineral P-54mg, Ca-470mg, I- 18.9mg

Economic value: The plant has got food value, as onion is used regularly as a vegetable or as a spice. The bulbs are eaten either as raw or in cooked form. It helps in reducing hypertension and has got anticancer value.

Allium cepa L.
Vernacular names
- Piaja (O)
- Onion (E)
- Pyaz (H)

5

Allium sativum L.

Hindi Name: Lahsun

English Name: Garlic

Family : Amaryllidaceae

Flowering: November-December

Fruiting: January-February

Botanical description: The plant is a herb. The leaves of the plants are flat, linear with sheaths, about half of their length. Flower of the plant is white or greenish white with a short pedicel. The plant is having bulbs with membranous outer scales in the axils of which do exist 10-12 lesser bulbs.

Ecology: The plant is commonly cultivated along the river banks during winter season. It is usually grows in sandy and loamy soils.

Use: Young leaves and shoots are collected, fried and eaten. Leaves in raw form are put in many curries and preparations. The bulbs are used as spices.

Nutritional Value: The plant is rich in minerals, vitamins and protein. The nutritional value of the plant is presented below. 100g of the plant contains about:

Carbohydrate 10.2g, Protein 4.55g, Fat 2.3g, Vitamin 16.32g

Mineral P-34mg, Ca-359mg, Fe- 15.9mg

Economic value: The plant has food value, as garlic is used regularly as a vegetable or as a spice. The plant has got medicinal value, used for blood purification and lowering blood pressure (hypertension).

Allium sativum L.
Vernacular names
- Rasuna (O)
- Garlic (E)
- Lahasun (H)

6

Allmania nodiflora R.Br. ex Wight

English Name: Node Flower Allmaina

Family: Amaranthaceae

Flowering: May-June

Fruiting: June-July

Botanical description: The plant is an annual herb. It is erect and rising, 10-50 cm tall. Stem is branched from or near the base. The Leaves are obovate, oblong, or linear with 1.5-6.5 mm long and 0.3-2.5 cm wide. Flowers of the plant are greenish or orange-red in colour with 3-7-flowered cymes. Seeds are single, black and shining with 1.5-2 mm in diam.

Ecology: The plant is a common weed widely distributed in forests and cultivated lands.

Use: Leaves and young shoots are roasted with mustard oil and eaten

Nutritional Value: The plant is rich in carbohydrate, vitamins and protein. The nutritional value of the plant is listed below. 1000g of the plant carries

Carbohydrate 17.6g

Fiber 1.8g

Protein 5.8g

Fat 0.4g

Vitamin 21.0g

Mineral P-40mg, Ca-420mg, I- 18.3mg

Allmania nodiflora R.Br. ex Wight
Vernacular names
- Chadheimundia saga (O)
- Node Flower Allmania (E)

7

Alternanthera sessilis L.R. Br.

Hindi Name: Garundi

English Name: Sessile Joyweed

Family: Amaranthaceae

Flowering: July-October

Fruiting: November-January

Botanical description: The plant is a perennial herb. The plant is a much branched prostrate herb, branches often purplish, frequently rooting at the lower nodes. The leaves are simple, opposite, somewhat fleshy, lanceolate, obtuse or subacute, glabrous and shortly petiolate. The flowers of the plant are small, white and in axillary clusters. Fruits are compressed obcordate utricles, with suborbicular seeds.

Ecology: The plant is often found in and near ponds, canals and water reservoirs. It prefers places with constant or periodically high humidity and are found in swamps, shallow ditches, waste lands and fallow rice fields.

Use: Leaves and young shoots are roasted with mustard oil and eaten. It is eaten with water rice.

Nutritional Value: The plant is rich in vitamins, proteins and minerals. The nutritional value of the plant is presented below. 100g of the plant contains:

Carbohydrate 11.6g, Fiber 2.8g, Protein 5.0g, Fat 0.7g, Vitamin 17.0g

Mineral P-60mg, Ca-510mg, Fe- 16.7mg

Economic value: The plant has medicinal value used against vomiting, diarrhea, dysentery, cough and snake bite. The plants are also eaten by cattles and buffalos.

Alternanthera sessilis L.R. Br.
Vernacular names
- Madaranga (O)
- Sessile Joyweed (E)
- Garundi (H)

8

Alternanthera ficoidea (L.) Sm.

Hindi Name: Kasul

English Name: Sanguinaria

Family : Amaranthaceae

Flowering: Most part of the year

Fruiting: Year round

Botanical description: The plant is an evergreen creeper, herbaceous and perennial, growing up to 15 – 30 cm in height and spreading to form a growth mat about 45cm wide or more. Stems are prostrate or erect and moderately branched, hairless below, generally hairy above and in leaf-axils. The leaves are inverted-lanceolate or narrowly elliptic, acute at apex, narrowed at the base into a poorly defined petiole, mostly 2-6 cm long and 0.6-2.0 cm wide. Round or short-cylindrical flower heads lie stalkless in leaf axils, 5-10 mm long and 5-7 mm wide.

Use: Young leaves and shoots are collected, fried and eaten as green vegetables

Ecology: It is distributed throughout India and commonly found in waste places pastures, waysides, ditches of agricultural lands etc.

Nutritional Value: The plant is rich in vitamins, protein and carbohydrate. The nutritional value of the plant is described below. 100g of the plant carries:

Carbohydrate 10.3g, Protein 5.8g, Fat 0.4g, Vitamin 16.8g

Mineral P-54mg, Ca-484mg, Fe-28.6mg

Economic value: The plant has got medicinal value, used against diarrhea and dysentery and often consumed by cattles and buffalos.

Alternanthera ficoidea (L.) Sm.
Vernacular names
- Banamadaranga(O)
- Sanguinarea (E)
- Kusal (H)

9

Amaranthus blitum subsp. *oleraceus* (L.) Costea

Hindi Name: Kosela

English Name: Red amarnath

Family: Amaranthaceae

Flowering: November-December

Fruiting: January-March

Botanical description: The plant is a herb, reaches heights of over 1 meter tall with fleshy oval shaped leaves that are sometimes pointed at the tips. It is having a deep maroon colour in the center of the leaf with streaks of red, purple and green. The plant produces feathery purple, magenta or red flowers from the central stalk which is packed with edible seeds. The flower buds are edible, once they mature and become bushy.

Ecology: The plant is commonly used as leafy vegetables and cultivated throughout the country. The plant loves to grow in the sandy soil in winter season. It is also cultivated in left over potato fields.

Use: Leaves and young shoots are cut into small pieces, cooked with salt and chilly and then eaten. Sometimes the leaves and shoots are mixed with vegetables in cooking.

Nutritional Value: The plant is rich in Minerals (Ca and Fe), vitamins and protein. The nutritional value of the plant is given below. 100g of the plan carries approximately:

Carbohydrate 6.4g

Fibre 1.5g

Protein 6.7mg

Fat 0.1g

Vitamin 28.3g

Mineral Ca-569mg, Fe-84.28mg

Economic value: The plant because of its softness, color and flavor has a good market value.

Amaranthus blitum subsp. *oleraceus* (L.) Costea
Vernacular name

- Kosila (O)
- Red amaranth (E)

10

Amaranthus caudatus L

Hindi Name: Kanta Chaulai

English Name: Prickly Amaranth

Family: Amaranthaceae

Flowering: June-August

Fruiting: September-November

Botanical description: The plant is an annual erect herb that grows upto 1 meter in height. The stem is grooved and often redish. The leaves are long-petioled, ovate or elliptic, acute or subacute with base shortly cuneate to attenuate. The flowers are in axillary and terminal red or green panicled spikes with up to 20 cm long. Seeds are black in colour, shining and compressed or subspherical with thick margin.

Ecology: The plant being commonly used as leafy vegetables, is cultivated throughout the country. The plant loves to grow in soil rich with organic materials. The plant is commonly cultivated in summer and it grows luxuriantly in rainy season.

Use: Leaves and young shoots are cut into small pieces, roasted with oil and eaten. Often the stems are cooked in curd and chillies.

Nutritional Value: The plant is rich in fibers, vitamins and protein. The nutritional value of the plant is presented below. The plant is a rich source of calcium.100g of the plant contains about:

Carbohydrate 7.8g, Fiber 9.35g, Protein 4.9mg, Fat 0.5g, Vitamin 24.7g

Mineral Ca-476mg, Fe-11.3mg

Economic value: The plant has a strong market value because of its taste, flavor and richness in fiber content.

Amaranthus caudatus L
Vernacular names:
- Khada (O)
- Prickly Amaranth (E)
- Kanta chaulai (H)

11

Amaranthus spinosus L.

Hindi Name: Kanta Chaulai

English Name: Prickly Amaranth

Family: Amaranthaceae

Flowering: June-August

Fruiting: September-December

Botanical description: The plant is an annual herb with sometimes red tinged erect stems, sometimes ascending, 30-150 cm long, usually branched. The leaves are ovate to rhombic- ovate, elliptic, lanceolate-oblong, or lanceolate, blades 1-12 cm long, 0.89-6 cm wide and smooth. The stem and its branches carry spines. Flowers of the plant are green, and are in axillary clusters in the lower part of the plant and in unbranched or branched spikes in the upper part. The lower clusters are entirely without stamens as the lower flowers of the spikes, the upper flowers in the spikes are staminate. Seeds are dark brown or black in colour.

Ecology: The plant is very common in waste grounds, roadsides, cultivated lands and distributed throughout India. The plant loves to grow in dry localities.

Use: Leaves and young shoots are cut into small pieces, cooked with salt and chilly and then eaten.

Nutritional Value: The plant is rich in vitamins and fibers and minerals such as calcium and iron. The nutritional value of the plant is described below. 100g of the plant contains approximately:

Carbohydrate 8.2g, Fiber 5.25g, Protein 6.6mg, Fat 0.7g, Vitamin 32.3g

Mineral Ca-845mg, Fe-33.6mg

Economic value: The plant is used as medicine aganist diarrhea and toothache.

Amaranthus spinosus L.
Vernacular names:
- Kanta Leutia (O)
- Prickly Amaranth (E)
- Kanta Chaulai (H)

12

Amaranthus tricolor L.

Hindi Name: Lal Khada

English Name: Elephant -Head Amaranth

Family: Amaranthaceae

Flowering: June-August

Fruiting: September-December

Botanical descripion: The plant is an erect tropical annual herb with long-stalked leaves, growing up to 5 ft tall. The leaves grow up to 5 inch long, are notched or rounded at the tips. The flowers are whitish-green or red white, the seeds are very small, shining, black or red- brown in colour.

Ecology: The plant is commonly cultivated as vegetable and often self shown in the tropical countries including India. The plant loves to grow in soil rich with compost/organic matter. It is cultivated in wet season.

Use: Leaves are collected, fried and eaten. Stems are often cooked with curd and chillies, that adds to its flavor.

Nutritional Value: The plant is rich in vitamins and proteins. The nutritional value of the plant is described below. 100g of the plant contains:

Carbohydrate 6.9g

Protein 8.6mg

Fat 0.6g

Vitamin 28.7g

Mineral Ca-390mg, Fe-24.8mg

Economic value: The plant has got good market value because of its taste and flavor.

Amaranthus tricolor L
Vernacular names
- Lal Khada (O)
- Elephant-Head Amaranth (E)

13

Amaranthus viridis L.

Hindi Name: Jangali Chaulayi

English Name: Wild Amaranth

Family: Amaranthaceae

Flowering: Throughout the year

Fruiting: Throughout the year

Botanical description: The plant is an annual herb with erect stems growing upto 10-80 cm in height. The leaves are triangular-ovate to narrowly rhombic, 2-7 cm long, 1.5-5.5 cm wide, hairless, with a narrow and small narrow notch. The flowers of the plant are green, in slender, paniculate spikes, in leaf axils or at the end of branches. Fruit is nearly round, about 1.3-1.5 mm in size.

Ecology: The plant is a fairly common weed on waste ground and roadsides and distributed throughout India. The plant is also cultivated during summer season.

Use: Leaves and young shoots are collected, roasted and eaten

Nutritional Value: The plant is rich in vitamins, minerals and protein. The nutritional value of the plant is described below. 100g of the plant contains:

Carbohydrate 7.4g

Protein 5.6mg

Fat 0.3g

Vitamin 32.7g

Mineral Ca-356mg,

Fe-19.3mg

Economic value: The plant is used as medicines against dysentery and inflammations

Amaranthus viridis L.
Vernacular names
- Leutia (O)
- Wild amaranth (E)
- Jungali chaulayi (H)

14

Andrographis paniculata (Burm.f.) Nees

Hindi Name: Kariyat

English Name: Creat

Family: Acanthaceae

Flowering: September-December

Fruiting: January-April

Botanical description: The plant is an erect annual herb extremely bitter in taste (all parts of the plant). It grows erect to a height of 1-4 ft. with dark stem. The leaves are glabrous, lanceolate, pinnate, with 8.0 cm long and 2.5 cm broad. Flowers of the plant are small, white-purple or spotted purple in colour, spreading axillary and terminal racemes or panicles and acute at both ends. Seeds numerous, sub quadrate and yellowish brown in colour.

Ecology: The plant is commonly found in forest, waysides and moist shady places. It is distributed throughout India

Use: Leaves are collected, fried with vegetables and eaten

Economic value: The plant is used as medicine against dysentery, cholera, pneumonia, chiken pox etc. The plant also acts to cure skin diseases and hypertension.

Andrographis paniculata (Burm.f.) Nees
Vernacular names:

- Bhuin nimba (O)
- Creat (E)
- Kariyat (H)

15

Arisaema tortuosum (Wall.) Schott

Hindi Name: Bagh Jandhra

English Name: Whipcord Cobra Lily

Family: Araceae

Flowering: July-August

Fruiting: November

Botanical description: The plant is a herb, growing about 90 cm in height. It produces 2 - 3 leaves, 5 - 30cm or more in size with a flowering stem 30 - 50cm long. Leaflets are lanceolate, acuminate and oblique, uniting in an intramarginal nerve. Fruiting head 5 cm long, ovoid, berries are initially green clustered fruits finally ripening to bright red.

Ecology: It is frequently found in open hill forests and also in moist shady places in forests and seen distributed throughout India.

Use: Young shoot/Fruit stalk is fried to prepare curry and eaten.

Economic value: Root/Tuber of the plant is used as medicine against rheumatism and joint pain.

Arisaema tortuosum (Wall.) Schott
Vernacular names:
- Bananada (O)
- Whipcord Cobra Lily (E)
- Bagh Jandhra (H)

16

Azadirachta indica A.Juss

Hindi Name: Neem

English Name: Neem

Family: Meliaceae

Flowering: February-May

Fruiting: June-July

Botanical description: The plant is a tree. The leaf of the plant is about 20-38 cm in length, leaflets 5-9 pairs and obliquely lanceolate. Flower is white in colour and fragrant. The plant bark, leaves and flowers are bitter in taste.

Ecology: It is very commonly planted, and often self shown. Also seen in forests as well as in wild. It is cultivated and naturalized throughout India.

Use: Fresh tender leaves along with flowers are fried with mustard oil. Often vegetables such as brinjal or potato is added during frying.

Nutritional Value: The plant rich in protein and minerals.

Protein 16.6 g

Mineral Ca-18.3 mg,

P-2.5mg, Mg-3.1mg

Economic value: The plant has strong medicinal value, used as medicine for worm infection and skin diseases. The dried leaves are used to protect seeds from insect infestation because of pesticidal value. The plant extracts/leaf powder is used as biopesticides. The plant/leaf extract is used in tooth paste preparation. The plant is used against hypertention and diabetics.

Azadirachta indica A. Juss
Vernacular names
- Nimba (O)
- Neem (E)
- Neem (H)

17

Bambusa bambos (L.) Voss

Hindi Name: Baans

English Name: Indian Thorny Bamboo

Family: Poaceae

Flowering & Fruiting: The plant flowers after about 90-100 years. Flowering and fruting occurs in wet season

Botanical description: The plant is a perennial tree growing up to 20-53 m in height and 8-18 cm in diameter. There are 1-3 spines at each branch node. Leaves are thin, linear, up to 20 cm long, 1-1.8 cm wide with a long pointed tip. The branches of the plant spread out from the base and areal roots reaching up to nodes. The length of the internodes is about 15-46 cm and is hollow.

Ecology: It is cultivated and naturalized throughout India. Plant loves to grow in moist soils.

Use: Young shoots (Karada) are cut into small pieces, cooked with salt and chilly and then eaten. Pickles are also prepared from young shoots in North Eastern States. The shoots (Karada) are also sold in tribal and rural markets by the local people.

Nutritional Value: The plant rich in protein and minerals

Economic value: Besides the nutritive value of the shoots, the plants are used for fencing and preparation of different containers. The leaves are used as fodder for cattles and other herbivores.

Bambusa bambos (L.) Voss
Vernacular names:
- Kanta baunsa (O)
- Indian Thorny Bamboo (E)
- Baans (H)

18

Basella alba L.

Hindi Name: Poi

English Name: Indian Spinach

Family: Basillaceae

Flowering: May-September

Fruiting: October-February

Botanical description: The plant is a twining succulent herb with green stem and petioles. Leaves of the plant are broadly ovate and green to purplish in colour. Flowers are pinkish or purple in colour. Fruits are berries, ovoid or depressed-globose, turns black after ripening with deep purple juice.

Ecology: It is commonly cultivated and often found in wild state on rocky grassy hill sides. Also widely cultivated throughout India during summer and rainy season.

Use: Stem and leaves are used to prepare curry. Preparations with fish is very tastey.

Nutritional Value: The plant is rich in minerals and vitamins. The nutritional value of the plant is presented below. 100g of the plant contains approximately:

Carbohydrate 3.40g

Protein 1.80g

Fat 0.30g

Vitamin A(8000IU), C(102mg), Niasin(0.500mg), riboflarin(0.155mg)

Mineral Ca(109mg), Fe(1.02mg), Mg(65mg)

Economic value: The plant has a good market value as different dishes are prepared from it. It is used against skin itching especially when caterpillar comes in contact.

Basella alba L.
Vernacular names:
- Poi (O)
- Indian spinach (E)
- Poi (H)

19

Bauhinia purpurea L.

Hindi Name: Kuilari

English Name: Geranium Tree

Family: Caesalpiniaceae

Flowering: September-February

Fruiting: February-March

Botanical description: The plant is a moderate sized tree with ashy or dark brown bark. Leaves are often oblong, about 5-18cm, deeply lobed from one third way down to nearly the base with scarcely overlapping. Flowers are purple and large on terminal panicled racemes. Pods are 15-30×18-25 cm long, narrow below, somewhat broadening upwards.

Ecology: It is frequent in forests and often planted. It is distributed in foothill areas also and cultivated throughout India.

Use: Young shoots along with leaves are collected, cooked as curry or fried and taken.

Nutritional Value: The plant is rich in carbohydrate and protein. The nutritional value of the plant is described below. 100 g of the plant contains about:

Carbohydrate 66.82g

Protein 15.19g

Fat 4.15g

Vitamin C(4.5mg)

Mineral Ca(240mg), Fe(21.73mg)

Economic value: The leaves of the plant are used as medicine against dysentery and skin diseases.

Bauhinia purpurea L.
Vernacular names:
- Kuilari/Barada (O)
- Geranium Tree (E)

20

Bauhinia vahlii wight & Arn.

Hindi Name: Malghan

English Name: Camel's Food Climber

Family: Caesalpiniaceae

Flowering: April-June

Fruiting: December-March

Botanical description: The plant is a climber growing up to 10-30 m long. The woody stem can get as thick as 20 cm. The spreading tout branches are covered with rusty fine hairs. The stout tendrils are coiling and occur in pairs. Large leaves are 10-45 cm, 2-lobed with a broad cut. The flowers are white, 2-3 cm across, turn yellow when old and borne in rounded clusters. Fruit is a flat woody pod with fine rusty hairs, 20-30 cm long.

Ecology: It is fairly common in mixed and sal forests widely distributed in the tropical countries including India.

Use: Young leaves are collected and cooked as curry and taken. The leaves are rich in protein. Also the seeds are eaten after boiling or cooked as vegetable.

Economic value: Plates are prepared from the leaves for selling in market.

Bauhinia vahlii wight & Arn.
Vernacular names:
- Siali (O)
- Camel's Foot Climber (E)
- Malghan (H)

21

Bauhinia variegata L

Hindi Name: Khairwal

English Name: Kachnar

Family: Caesalpiniaceae

Flowering: February-March

Fruiting: April-May

Botanical description: The plant is a moderate sized tree growing up to 20-40 ft tall and 10-20 ft wide with a spreading crown of briefly deciduous leaves which are 10-15 cm across and rounded with lobed ends and heart shaped bases. The leaves are shaped like a cow's hoof. The flowers are white or purple in colour with one petal variegated yellow, 5-10 cm across and carried in clusters at the branch tips. The flowers often make their first appearance in late winter while the tree is bare of leaves. Seeds of the plant are sub-orbicular in shape.

Ecology: It is commonly found in mixed forests, especially in the hills, often planted in gardens and distributed throughout India.

Use: Young leaves are collected, cooked as curry and taken. The flowers coated with rice powder are very good in taste after fried.

Nutritional Value: The plant is rich in vitamin C and other trace elements such as calcium and magnesium.

Economic value: The plant is used as medicine against diarrhoea and worm infested wounds.

***B**auhinia variegata* L
Vernacular names
- Kanchan (O)
- Kachnar (E)
- Khairwal (H)

22

Benincasa hispida (Thunb.) Cogn.

Hindi Name: Petha
English Name: White Gourd

Family: Cucurbitaceae

Flowering: October-November

Fruiting: December-January

Botanical description: The plant is a annual climbing herb, having stems up to 6 metres long that scramble over the ground or climb on other plants, supporting themselves by means of tendrils. Leaves are roundish, kidney-shaped, base deeply heart-shaped. Upper surface is rough, lower surface shortly bristly, 5-7-lobed, with long and hairy stalk. Tendrils are slender and rarely simple. Flowers are yellow in colour, calyx tube 10-15 mm long with spreading petals ending in a short point. Fruits large, succulent, densely hairy when young, with a thick waxy deposit when mature.

Ecology: It is widely cultivated as vegetable throughout India especially in wet season. The plant loves to grow in humus reach soil in the wild.

Use: Tender leaves are collected, cooked as curry or fried and then eaten.

Economic value: Fruit of the plant is commonly used as a vegetable in mixed curry preparation. Special dishes 'Badi' (dried ball) is also prepared from it. Seeds are fried and eaten.

Benincasa hispida (Thunb.) Cogn.
Vernacular names
- Panikakharu (O)
- White Gourd (E)
- Petha (H)

23

Boerhavia diffusa L.

Hindi Name: Punarnava

English Name: Hog Weed

Family: Nyctaginaceae

Flowering & Fruiting: Most parts of the year, mostly dry winter season

Botanical description: The plant is a prostrate herb with diffuse inflorescences. The stem of the plant is glabrous or minutely pubescent and often purplish. The leaves are in unequal pairs, broadly ovate with obtuse or rounded apex. Flowers are very small, purplish red to reddish pink or nearly white, bell-shaped beyond the constriction.

Ecology: It is a common weed found near dry and warm river valleys; along roadsides, near habitations, in and along cultivated fields and in open cleared patches in forests. It is distributed throughout India.

Use: Tender young leaves and shoots are cooked and eaten.

Nutritional Value: The plant is rich in minerals and proteins. The nutritional value of the plant is presented below. 100g of the plant carries approximately:

Carbohydrate 10.56g

Protein 12.4g

Fat 2.5g

Vitamin Ca (41.60mg), B3(66.20mg)

Mineral Na(75.55mg), Ca(204.00mg), Mg(6.63mg)

Economic value: The plant is having medicinal value and used for the treatment of jaundice and gastric disturbances.

Boerhavia diffusa L.
Vernacular names
- Puruni saga (O)
- Hog weed (E)
- Punarnava (H)

24

Brassica napus L.

Hindi Name: Sarshon
English Name: Rape

Family: *Brassicaceae*

Flowering: September-November

Fruiting: December-February

Botanical description: The plant is an annual herb with stems reaching up to 1.5 m tall. Stems are erect, reddish-purple below, greenish-red above, hairless, arising from large thickened taproot up to 10 cm in diameter. Leaves are large alternately arranged and hairless, upper smaller. Lower leaves are up to 25 cm long, pinnately cut, with small prickles above and below. Flowers are yellow in colour, borne in racemes at the end of branches with 2 cm across. The pods are stout, 5-9 cm long bearing 30-80 seeds with whitish, yellow or brownish in colour.

Use: Leaves are fried with oil then eaten and also chopped leaves and stems are dried in shade, stored in an earthen pot for consumption.

Ecology: It is cultivated and often found as an escape. It is widely cultivated in India.

Nutritional Value: The plant is rich in vitamins and minerals. The nutritional value of the plant is presented below. 100g of the plant carries:

Carbohydrate 4.9g

Protein 2.7g

Fat 0.2g

Vitamin A-43.3mg, B1-0.56mg, B2-

0.158mg, C-26.4mg,

Mineral P-21mg, Ca-49mg, I-0.33mg,

Mg-14mg

Economic value: The plant is commonly cultivated, as the seeds are used as one of the major spices. Edible oil is extracted from the seed(s).

Brassica napus L.
Vernacular names
- Sorisa saga (O)
- Rape (E)
- Sarshon (H)

25

Brassica oleracea L. var *botrytis* L.

Hindi Name: Phulgobi

English Name: Cauliflower

Family: Brassicaseae

Flowering: November-February **Fruiting:** February-March

Botanical description: The plant is a herb with white edible mass of abortive flowers aggregated on thick branches of the inflorescence.

Ecology: It is widely cultivated as vegetable in the tropical countries including India. The plant grows in winter season.

Use: Leaves are fried or made into curry and consumed. Inflorescence is used in various types of preparations.

Nutritional Value: The plant is rich in minerals and vitamine A and B complex. The nutritional value of the plant is described below. 100g of the plant contains approximately:

Carbohydrate 6.8g

Protein 3.4g

Fat 0.2g

Vitamin A-38.4mg, B1-0.51mg, B2- 0.28mg, C-26.8mg, Mineral P-30mg, Ca-32mg, Fe-12.6mg, Mg- 10mg

Economic value: The plant is commonly cultivated and used as a vegetable. The leaves are also used as cattle feed.

Brassica oleracea L. var *botrytis* L.
Vernacular names
- Phulkobi (O)
- Cauliflower (E)
- Phulgobi (H)

26

Brassica oleracea L. var *capitata* L.

Hindi Name: Bandgobi

English Name: Cabage

Family: Brassicaseae

Botanical description: The leaves of the plant is having stout nerves, densely congested into a subglobose or ellipsoid head which is edible.

Ecology: The plant is a herb, widely cultivated as vegetable in the tropical countries including India.

Use: Leaves are fried separately or with other vegetables, and then eaten

Nutritional Value: The plant is rich in vitamin B and minerals.The nutritional value of the plant is presented below. 100g of the plant leaves carries:

Carbohydrate 5.8g

Protein 1.3g

Fat 0.1g

Vitamin A-41.3mg, B1-0.061mg, B2-0.140mg, C-36.6mg,

Mineral P-26mg, Ca-40mg, Fe-0.47mg, Mg- 12mg

Economic value: The plant is commonly cultivated and used as a popular vegetable

Brassica oleracea L. var *capitata* L.
Vernacular names
- Patrakobi/Bandhakobi (O)
- Cabbage (E)
- Bandgobi (H)

27

Brassica oleracea L. var. *gongylodes* L.

Hindi Name: Ganth-gobi

English Name: Knol-khol

Family: Brassicaseae

Flowering & Fruiting: August-March

Botanical description: The plant is having a short stem, enlarging into a spherical edible tuber. Basal cauline leaves are broadly ovate, not overlapping on globose stem, about 10-45 x 3-25 cm across with cuneate base.

Ecology: The plant is a herb, widely cultivated as vegetable in the tropical countries including India. The plant loves to grow in late winter season.

Use: Leaves are fried separately or with other vegetables and then eaten. Curries are preparaed with patato

Nutritional Value: The plant is rich in vitamin B, C and minerals. The nutritional value of the plant is presented below. 100g of the plant contains about:

Carbohydrate 3.2g

Protein 0.89g

Fat 0.2g

Vitamin A-3.2mg, B_1-0.041mg, B_2- 0.110mg, C-28.4mg,

Mineral P-22mg, Ca-35mg, Fe-0.18mg, Mg- 5mg

Economic value: The plant is commonly cultivated and used as a vegetable.

Brassica oleracea L. var. *gongylodes* L.
Vernacular names
- Ganthi kobi/Olkobi (O)
- Knol-khol (E)
- Ganth Gobi (H)

28

Celosia argentea L.

Hindi Name: Kurdu

English Name: Quail grass

Family: Amaranthaceae

Flowering: August-October

Fruiting: November-January

Botanical description: The plant is a herb, growing up to 60-75cm height, erect, branching with oval or lanceo-shaped, with hundreds of tiny flowers packed in dense spikes of silver- white flowers which usually stand above the foliage. They are beautiful plants with soft, dense feathery spikes, produced in profusion. Leaves are lanceolate, linear, acute to acuminate tapering into slender petiole. Seeds are several, black in colour and shining.

Ecology: The plant is generally seen in open moist places, hills, in arable lands, and as a weed of cultivation and in harvested fields.

Use: Young leaves and shoots are collected, fried then eaten.

Economic value: The plant is having medicinal importance and used against skin eruptions

Celosia argentea L.
Vernacular names
- Ghurudi saga (O)
- Quail grass (E)
- Kurdu (H)

29

Centella asiatica L.

Hindi Name: Ballari

English Name: Indian Pennywort

Family: Apiaceae

Flowering: August-October

Fruiting: November-January

Botanical description: The plant is a small creeping herb with shovel shaped leaves emerging alternately in clusters at the stem nodes. The runners lie along the ground and the inch long leaves with their scalloped edges rise above on long reddish petioles. The insignificant greenish- to pinkish-white flowers are borne in dense umbels (clusters in which all the flower stalks arise from the same point) on separate stems in the summer. The seeds are pumpkin-shaped nutlets 0.1-0.2 cm long.

Ecology: It is commonly seen in shady, damp and wet places such as paddy fields and in grass thickets and distributed throughout India.

Use: Leaves and young shoots are collected, roasted then eaten

Nutritional Value: The plant is a rich source of protein and calcium. The nutritional value of the plant is described below. 100g of the plant contains about:

Carbohydrate 8.6g

Protein 10.00g

Fat 0.27g

Vitamin C-2.93g, B_1-57mg

Mineral Ca-453mg, Fe-682mg, Mg 0.561mg

Economic value: The plant is used as medicine for stomach disorder and asthma

Centella asiatica L.
Vernacular names
- Thalkudi (O)
- Indian Pennywort (E)
- Ballari (H)

30

Chenopodium album L.

Hindi Name: Bathuwa

English Name: White Goosefoot

Family: Chenopodiaceae

Flowering: November-January **Fruiting:** February-April

Botanical description: The plant is a fast-growing, upright and weedy annual herb. The leaves are variables in appearance. The first leaves, near the base of the plant, are toothed and 3-7 cm long and 3-6 cm broad. The leaves on the upper part of the flowering stems are entire and lanceolate-rhomboid with 1-5 cm long and 0.4-2 cm broad. The flowers of the plant are 1-1.2 mm across, in short axillary spikes and often forming a terminal thyrse. Seeds are black, depressed with an obscure horizontal keel.

Ecology: It is a common weed in wasteland and cultivated ground, especially on rich soils and old manure heaps and distributed throughout India.

Use: Leaves and young shoots are cooked along with other vegetables by adding salt and chilly to it.

Nutritional Value: The plant is rich in Vitamins and mineals. The nutritional value of the plant is presented below. 100g of the plant contains approximately:

Carbohydrate 7.3g

Protein 4.2g

Fat 0.8g

Vitamin C(80mg), B6(0.274mg), B3(1.2mg), B1(0.16mg), B2(0.04mg)

Mineral Ca(309 mg), Fe (1.2mg),Mg(34mg), Mn(0.782mg)

Economic value: The plant is used as medicine to cure blood dysentery.

Chenopodium album L.
Vernacular names
- Bathua saga (O)
- White Goosefoot (E)
- Bathuwa (H)

31

Cicer arietinum L.

Hindi Name: Chana

English Name: Chickpea

Family: Fabaceae

Flowering: December-February

Fruiting: February-April

Botanical description: The plant is an annual herb, much branched and viscous hairy. The plant grows up to 20-50 cm high and has small feathery leaves on either side of the stem. The flowers are bluish purple in colour, 6-7 mm long on slender peduncles. Pods oblong, one seedpod contains two or three peas.

Ecology: It is commonly cultivated as a cold weather crop throughout India.

Use: Young leaves and shoots are collected, fried then eaten.

Nutritional Value: The plant is rice in carbohydrate and proteins. 100g of the plant contains about:

Carbohydrate 8.93g

Protein 3.9g

Fat 0.76g

Economic value: The plant is also a major grain legume, widely cultivated for its edible seeds consumed as raw or cooked.

Cicer arietinum L.
Vernacular names
- Buta saga (O)
- Chickpea (E)
- Chana (H)

32

Cinnamomum tamala Buch. Ham.

Hindi Name: Tej Pata

English Name: Indian bay leaves/Indian Cassia

Family: Lauraceae

Flowering: January **Fruiting:** February

Botanical description: The plant is a moderate sized tree growing up to 10-20 metres. Leaves are mostely oblong, usually 10-12.5 cm long, acuminate or acute, glabrous with shining above and acute base. The flowers are 5-6 mm long with pubescent pedicel. The tepals are pubescent, deciduous in fruits.

Ecology: The plant is occasionally planted and commonly cultivated as vegetable in the tropical countries including India.

Use: Leaves are used in different curries and recipes for special aroma.

Nutritional Value: The plant is rich in minerals and vitamins B complex. The nutritional value of the plant is presented below. 100g of the plant contains approximately:

Carbohydrate 74.97g

Protein 7.61g

Fat 8.36g

Vitamin B_2 (0.421mg), B_3 (2.005mg)

Mineral Ca (834mg), Cu (0.416mg), Fe (43mg), Mg (120mg), Zn (3.70mg)

Economic value: The plant has got medicinal value and used against cough and cold.

Cinnamomum tamala Buch. Ham
Vernacular names
- Teja patra (O)
- Indian bay leaves/Indian Cassia (E)
- Tej pata (H)

33

Cleome gynandra L.

Hindi Name: Safed hulhul

English Name: Wild Spider Flower

Family: Cleomaceae

Flowering: July-October

Fruiting: December-February

Botanical description: The plant is an erect herbaceous annual herb growing up to 1.5 m in height. Leaves are alternately arranged and stalked. Each leaf has 3-7 leaflets, which are pinnately dissected and stalkless. Inflorescence is quite showy, and is usually up to 30 cm in length. The flowers have long stalks, ovate or elliptic petals either purplish or white in colour. The fruit is a long-stalked, dry, spindle-shaped capsule. Seeds are many, 1.5 mm across, with a shallow and narrow cleft.

Ecology: It is a frequently seen weed in open areas, waste places, cultivated and uncultivated land, and distributed throughout the warmer parts of India.

Use: Leaves and young shoots are collected, roasted then eaten

Nutritional Value: The plant is rich in minerals, vitamins and protein. The nutritional value of the plant is given below. 100g of the plant contains:

Carbohydrate 6.2g

Protein 7.6g

Fat 0.4g

Vitamin B1-0.843mg, C-964mg

Mineral Ca(746mg), P(69mg), Fe(32mg),

Economic value: The plant is used as medicine for the treatment of scorpion stings

Cleome gynandra L.
Vernacular names
- Araka saga (O)
- Wild Spider Flower (E)
- Safed hulhul (H)

34

Cleome viscosa L.

Hindi Name: Bagra

English Name: Asian Spider Flower

Family: Cleomaceae

Flowering: May-July

Fruiting: August-October

Botanical description: The plant is a tall annual herb growing up to a meter high, more or less hairy with glandular hairs. The leaves are compound, with 3-5 leaflets. Leaflets are obovate, elliptic-oblong, very variable in size. Flowers are yellow in colour, 10-15 mm across with 6-20 mm long pedicels. Fruits are 30-75 mm long, 3-5 mm broad, linear-oblong, erect, with tapering at both ends. Seeds many, 1.25mm across with dark brown colour.

Ecology: The plant is commonly grown as a weed in fallow lands, fields, roadsides and wastelands. The plant loves to grow in sandy and freely draining soils in open woodland scrub and on scree slopes in dry areas. It is distributed throughout the warmer parts of India.

Use: Leaves and young shoots are collected, fried/ roasted then eaten

Nutritional Value: The plant is rich in minerals and proteins. The nutritional value of the plant is presented below. 100g of the plant contains approximately:

Carbohydrate 7.8g

Protein 5.6g

Fat 1.2g

Vitamin B1-0.985mg, C-564mg

Mineral Ca(880mg), P(73mg), Fe(24mg),

Economic value: The plant is used as medicine for digestive disorders. Seeds are rich in oil.

Cleome viscosa L.
Vernacular names
- Bana sorisa (O)
- Asian Spider Flower (E)
- Bagra (H)

35

Colocasia esculenta (L.) Schott

Hindi Name: Kachu

English Name: Elephant Ear Taro

Family : Araceae

Flowering: The plants flowers in Late wet season (September-October)

Botanical description: The plant is a tuberous bulb, herbaceous plant growing about 3-5 ft in height. The plant has large leaves. The leaves are heart shapes, 2-3 ft long with a long stalk of about 3 ft height. The flowers are yellowish-white spathes and spadixes are frequently produced.

Ecology: It is commonly found in damp, shady places and also cultivated. The plant is widely distributed throughout the hotter parts of India.

Use: Leaves are collected, fried with oil then eaten. Stalks are boiled and mixed with potato in preparation of curries.

Nutritional Value: The plant is rich in minerals and vitamins. The nutritional value of the plant is presented below. The plant is rich in potassium and magnesium. 100g of the plant contains about:

Carbohydrate 5.8g

Protein 3.9g

Fat 0.6g

Vitamin A-6145, IU B_2-0.7mg, C-51.5mg

Mineral Ca-(125mg), P(39.2mg), Fe(1.7mg), Mg (29.0mg)

Economic value: The plant is commonly cultivated as tuber crop. The tuber is used in many type of curries and preparations.

Colocasia esculenta (L.) Schott
Vernacular names
- Saru(O)
- Elephant Ear Taro (E)
- Kachu/Kachlu (H)

36

Commelina attenuata K.D.Koenig ex Vahl

Hindi Name: Kanseera

English Name: Asian Spider Flower

Family: Commelinaceae

Flowering: August-September

Fruiting: September-October

Botanical description: The plant is a small diffused or straggling herb. Stem usually with a line of pubescence. Leaves are linear, 3.7-7.5 cm long, acute or acuminate, glabrous with prominent midrib and acute-obtuse at apex. Flowers are blue in colour and very small. Capsules are oblong and 4.3 mm long. The capsule is having 2 seeds, 2.5 mm long, ellipsoid, cylindrical with white appendage at each end.

Ecology: It is frequent in moist places, sandy soils or rocky slopes and also in moist or wet thickets or forest and distributed throughout India.

Use: Leaves and young shoots are collected, fried/ roasted and eaten.

Nutritional Value: The plant is rich in vitamins and minerals. The nutritional value of the plant is presented below. The plant is a rich source of calcium, potassium and iron.100g of the plant contains approximately:

Carbohydrate 5.2g

Protein 4.6g

Fat 0.4g

Vitamin: B1-0.56mg, C-62mg, E- 10.2mg

Mineral: Ca(652mg), P(68mg), Fe(21mg),

Economic value: The plants and the seeds being rich in protein are used on cattle feeds.

Commelina attenuata K.D.Koenig ex Vahl
Vernacular names
- Chhena saga (O)
- Asian Spider Flower (E)
- Kanseera (H)

37

Commelina benghalensis L.

Hindi Name: Kana

English Name: Bengalday Flower

Family: Commelinaceae

Flowering: July-October

Fruiting: November-January

Botanical description: The plant is a diffuse, branched, sub-succulent herb growing upto 30-60 cm. Leaves are 3-5 x 2-3.5 cm in size, ovate or elliptic-ovate, base rounded or almost flat with blunt or pointed tip. Spathe is about 1.5 cm across, funnel-shaped, flat at apex and velvet-hairy. Sepals are nearly equal, about 2.5 mm long, outer ones linear, inner ones orbicular. Flowers are small, petals are blue in colour, larger ones about 4 x 4.5 mm with broadly ovate. The seeds are 5 and closely pitted.

Ecology: It is a very common weed found in waste lands and waysides, mostely in wetlands, distributed throughout India. The plant loves to grow in rice fields after harvest.

Use: Leaves and young shoots are collected, fried/ roasted and eaten

Nutritional Value: The plant is rich in minerals and vitamins. The nutritional value of the plant is presented below. The plant carries high content of calcium, magnesium, potassium and phosphorous. 100g of the plant carries:

Carbohydrate 4.69g

Protein 3.98g

Fat 0.64g

Vitamin C(39mg), E(6.5mg), B_1(0.076mg). Mineral Ca(1431.6mg), Mg(220.8mg) N(500mg), P(192mg), K(390mg),

Economic value: The plant is also used as cattle feed.

Commelina benghalensis L
Vernacular names
- Kansiri (O)
- Bengal Dayflower (E)
- Kana (H)

38

Corchorus capsularis L.

Hindi Name: Titapat

English Name: White jute

Family: Tiliaceae

Flowering: July-September

Fruiting: October-November

Botanical description: The plant is an erect, annual herb reaching up to 1-2 m height. The stems are usually purplish white. Leaves are ovate-lanceolate, 5-12 cm long, pointed at the tip and rounded at the base. The flowers are borne in small groups in the axils of the leaves about 4 mm long. The capsules are round to ovoid and about 1 cm in diameter.

Ecology: It is commonly cultivated throughout India for fiber purpose. The plant loves to grow in wet soil.

Use: Tender leaves are collected, curry is prepared and eaten.

Nutritional Value: The plant is rich in Vitamin B and minerals. The nutritional value of the plant is listed below. 100g of the plant contains:

Carbohydrate 12.4g

Protein 5.6g

Fat 0.28g

Vitamin B_1(15mg), B_2(28mg), B_3(1.5mg), C (95mg)

Mineral Ca(366mg), P(122mg), Fe(11.6mg),

Na(12mg), k(444mg)

Economic value: The fibre is extracted from the stems after retting in water. The fibres are used for preparing of gunny bags, carry bags, ropes etc.

Corchorus capsularis L.
Vernacular names
- Nalita (O)
- White jute (E)
- Titapat (H)

39

Coriandrum sativum L.

Hindi Name: Dhania
English Name: Coriander

Family: Apiaceae

Flowering: December-January

Fruiting: February-March

Botanical description: The plant is a herb, growing to a height of about 2 feet with a spread of 9 inches. The leaves are bright green in colour, fan shaped and become more feathery towards the top of the plant. The flowers are small and white, formed in umbel-like clusters. Fruit is globose, glabrous and seeds are flattened dorsally in transaction.

Ecology: The plant is commonly cultivated throughout India, and sometimes found in waste places and arable land, often by the sides of rivers. It loves to grow in sandy and loamy soils. The plant loves to grow in cold season.

Use: Leaves and shoots are collected, added to curry for aroma and also chutney is prepared from the fresh leaves.

Nutritional Value: The plant is rich in vitamin C and B complex. The nutritional value of the plant is described below. 100g of the plants contains:

Carbohydrate 3.67g

Protein 2.13g

Fat 0.52g

Vitamin: C(27mg), E (2.5mg), B_1(0.006 mg), B_2 (0.16mg), B_3 (1.114mg) Beta carotene (3930 micro g)

Mineral: Ca(67mg), Fe(1.77mg), Mg(26mg)

Mn(0.426mg), Zn(0.5mg)

Economic value: Leaves are used as medicine for stomach disorder and help in digestion.

Coriandrum sativum L
Vernacular names
- Dhania (O)
- Coriander (E)
- Dhania (H)

40

Cucurbita maxima Duchesne.

Hindi Name: Lal Kaddu

English Name: Sweet Gourd

Family: Cucurbitaceae

Flowering: March-June

Fruiting: May-August

Botanical description: The plant generally spreads on the ground with cylindric and grooved stem. The leaves are rounded to reniform, 6-25cm in diameter with 5 rounded shallow lobes and coarsely hairly. The flowers are monoecious and yellow in colour. Fruits usually lagre, oblong, oval, flattened or globular in size. Seeds are ovate, obscurely margined with white or yellow in colour.

Ecology: It is an annual climbing herb, widely cultivated as vegetable in the tropical countries including India. The plant loves to grow in sandy and loamy areas.

Use: Leaves and stems are collected and cooked as curry by cutting into small pieces and eaten. Young male flowers are also eaten raw or cooked.

Nutritional Value: The plant is rich in vitamins and minerals. 100g of the plant carries about:

Carbohydrate 4.82g

Vitamin A(49.81), C(31.42)

Mineral P(84.9 mg)

Economic value: The fruit of the plant is having nutritional importance and commonly used as vegetable. The leaves and stems are used as cattle feed.

Cucurbita maxima Duchesne.
Vernacular names
- Kakharu saga (O)
- Sweet gourd (E)
- Lal Kaddu (H)

41

Dendrocalamus strictus (Roxb.) Nees

Hindi Name: Bans

English Name: Bamboo

Family : Gramineae (Poaceae)

Flowering: November

Botanical description: The plant is a perennial tree that grows up to 20-53 m in height and 8-18 cm in diameter. There are 1-3 spines at each branch node. Leaves are thin, linear, up to 20 cm long, 1-1.8 cm wide with a long pointed tip. The branches of the plant spread out from the base and areal roots reaching up to nodes. The length of the internodes is about 15-46 cm and is hollow.

Ecology: It is fairly common in mixed forests or dry hills, often found in hills and shady slopes and alluvial plains.

Use: Young shoots (Karada) are cut into small pieces, cooked with salt and chilly and eaten. Also pickles are prepared after adding spices and vinegar. The shoots (Karada) are also sold in tribal and rural markets by the local people.

Nutritional Value: The plant shoot is rich in protein and roughage.

Economic value: Bamboo is used for making houses and furnitures. The plant has also strong medicinal/pesticidal value. The leaves are eaten by cattles and buffalos.

Dendrocalamus strictus (Roxb.) Nees
Vernacular names
- Baunsa (O)
- Bamboo (E)
- Bans (H)

42

Emilia sonchifolia (L.) DC. ex DC.

Hindi Name: Hirankhuri

English Name: Purple Sow Thistle

Family: Compositae

Flowering: Aug-December

Fruiting: January-April

Botanical description: The plant is an annual herb. The stems are weak, erect or often branched at the base, smooth or sparingly hairy with 10 to 60 cm tall. This species is recognized by the sow-thistle like leaves. Lower leaves are deeply and irregularly toothed, kidney-shaped, ovate, triangular-ovate or obovate, 4-16 cm long, 1-8 cm wide with narrowly winged stalks. Upper leaves are smaller, alternately arranged, usually entire, sometimes coarsely toothed. Inflorescence is an involucrate flower head resembling a single flower, long-stalked, at the end of branches. The cup of the flower-head is green, cylindrical, somewhat inflated below. Florets are 30-60 per head, purple or pink in colour,

Ecology: It is an occasional common weed in moist areas and uncultivated grounds in moist, shady places, often found in cultivated lands and distributed throughout India. The plant usually grows in sandy and loamy soils.

Use: Tender leaves and young shoots are collected, cooked then eaten

Emilia sonchifolia (L.) DC. ex DC.
Vernacular names
- Musakani saga (O)
- Purple Sow Thistle (E)
- Hirankhuri (H)

43

Enydra fluctuans DC.

Hindi Name: Hira Kuch

English Name: Purple Sow Thistle

Family: Asteraceae

Flowering: December-January

Fruiting: January-February

Botanical description: The plant is a prostrate herb. The stem of the plant is succulent, with short ascending flowering hairy branches. The leaves are narrowly linear-oblong, acute or obtuse, entire or subcrenate with 2-5cm long. Heads yellow in colour, axillary and terminal, sessile with 7-18mm diameter.

Ecology: It is usually found in swamps and aquatic areas in coastal regions etc and occasional in wet places along river beds and ponds.

Use: Leaves are cooked with mustard oil, chilly, salt and eaten.

Nutritional Value: The plant is rich in minerals and carbohydrate. The nutritional value of the plant is presented below. The plant is a rich source of calcium and magnesium. 100g of the plant contains:

Carbohydrate 6.1g

Protein 1.6g

Fat 0.2g

Vitamin: C- 418mg, B_1-0.69mg

Mineral: Ca-902mg, Na-135mg, Mg-57mg, Fe-23mg

Enydra fluctuans DC.
Vernacular names
- Hidmicha (O)
- Water cress (E)
- Harakuch (H)

44

Eryngium foetidum L.

Hindi Name: Ban Dhania

English Name: Long Coriander

Family: Apiaceae

Flowering: April-June

Fruiting: July-August

Botanical description: The plant is an perennial, erect, dichotomously branched, aromatic herb, growing up to 15-45 cm high. Leaves are sessile, oblanceolate-oblong, spinosely- serrate and glabrous. Inflorescence heads numerous, cylindric, about 10 x 5 mm with the involucral bracts exceeding the heads. Petals are white or greenish in colour. Fruits are greenish, subspherical and about 1.5 mm in diameter.

Ecology: It is usually found in damp places close to streams in cool valleys, along roadsides and forest edges and distributed throughout India. Plant loves to grow on moist soil.

Use: Leaves and shoots are collected, added to curry for aroma and also chutney is prepared from the fresh leaves

Economic value: The plant having medicinal importance and used to cure fever, malaria, stomach pain etc.

Eryngium foetidum L.
Vernacular names
- Jangali dhania (O)
- Long Coriander (E)
- Ban dhania (H)

45

Furcraea foetida (L.) Haw.

English Name: Giant Cabuya

Family: Asparagaceae

Flowering: October-January

Botanical description: The plant is an evergreen perennial shrub. It is stemless or with a short stem up to 1 m tall, producing a rosette of leaves of about 40 - 180cm long. The leaves are sword-shaped, 1-1.8 m long and 10-15 cm broad at their widest point, narrowing to 6-7 cm broad at the leaf base, and to a sharp spine tip at the apex. The flowers are greenish to creamy white, 4 cm long, and strongly scented.

Ecology: It is usually planted in gardens and also found as a weed of bushland or open woodland and distributed throughout India.

Use: Young shoots are collected, fried, and then eaten.

Nutritional Value: The plant is rich in protein. It carries about 12.5g protein per 100g of shoots.

Economic value: The flower of the plant is having ornamental value as it is very scented.

Furcraea foetida (L.) Haw.
Vernacular names
• Muruga (O)
• Giant Cabuya (E)

46

Glinus oppositifolius (L.) A.D.C.

Hindi Name: Jima

English Name: Glinus

Family: Molluginaceae

Flowering: March-June

Fruiting: July-October

Botanical description: The plant is a prostrate annual herb, growing upto to 50 cm long. Leaves are in pseudo-whorls of 3-6 or opposite; leaf blade is spoon-shaped or elliptic, obtuse or acute apex with short leaf stalks. Flowers are greenish white in colour and about 5mm to 8mm across. Capsule is ellipsoid, shorter than the sepals. Seeds are brown and finely granulate.

Ecology: It is commonly found in sandy soil, especially alone river banks, occasionally found in swampy localities, in dried-up pools and ditches and in rice-fields. The plant loves to grow on sandy soil.

Use: Young leaves and shoots are collected, roasted with other vegetables and then eaten

Nutritional Value: The plant is rich in minerals, protein and carbohydrate. The nutritional value of the plant is presented below. The plant is a rich source of Zinc and iron. 100g of the plant carries:

Carbohydrate 38.2g

Protein 12.5g

Fat 2.3g

Vitamin: C(112mg), B_3(0.1mg),

Mineral: Ca(1693mg), Fe(22.1mg), Zn(43.2mg)

Economic value: The plant is used as medicine for itchs and other skin diseases.

Glinus oppositifolius (L.)A.DC.
Vernacular names
- Pita saga (O)
- Glinus (E)
- Jima (H)

47

Hibiscus sabdariffa L.

Hindi Name: Lal Ambari

English Name: Roselle

Family: Malvaceae

Flowering: July-October

Fruiting: October-February

Botanical description: The plant is a robust branched, annual herb that grows 4-7 ft tall and almost as broad. Leaves are dark green in colour, about 6 in across, deeply dissected into 3-5 narrow lobes. The stems, branches, leaf veins, and leaf stalks are reddish purple. The flowers of the plant are yellow with deep purple centre and about 3 inches across. The sepals of the flowers are prominent, fleshy, bright red in colour. Capsule is ovoid, 1.5 cm across with brownish seeds.

Ecology: It is cultivated and often self sown in disturbed lands.

Use: The leaves are very sour in taste. Leaves are collected, cooked with water and other vegetables to prepare sour water 'sambar'.

Nutritional Value: The plant is rich in vitamin C and minerals. The nutritional value of the plant is presented below. 100g of the plant contains:

Carbohydrate 6.45g

Protein 2.73g

Fat 0.36g

Vitamin C-6.8mg, B_1-0.006mg, B2-0.016mg,

Mineral Ca-123mg, Fe-0.84mg, Mg-29mg

Economic value: The plant is diuretic in nature and used as medicine for stomach and digestive disorders.

Hibiscus sabdariffa L.
Vernacular names
- Khatta palanga/Kaunria saga (O)
- Roselle (E)
- Lal Ambari (H)

48

Hybanthus enneaspermus (L.) F.Muell.

Hindi Name: Ratan Purush

English Name: Medan Mastak

Family: Violaceae

Flowering: Round the year

Fruiting: Round the year

Botanical description: The plant is a perennial herb The plant grows upto to 60 cm high, often branched and spreading. Leaves are sub-sessile, linear to lanceolate, 1-5 cm long with recurved to revolute margins or occasionally flat. The flowers are pink in colour with 5 petals. Capsule is subglobose, 4-9mm long with 5-12 seeds pitted between ribs.

Ecology: It is commonly found in lawns and cultivated lands and widely distributed throughout India.

Use: Leaves and young shoots are collected, fried/ roasted then eaten

Economic value: The plant has medicinal value, used to cure stomachache and dysentery.

Hybanthus enneaspermus (L.) F.Muell.
Vernacular names
• Luni saga/Madan mastak (O)
• Ratan Purush (H)

49

Hydrolea zeylanica (L.) Vahl

Hindi Name: Koilary

English Name: Ceylon Hydrolea

Family: Hydroleaceae

Flowering: November-January

Fruiting: February-April

Botanical description: The plant is an annual, aquatic, ascending or prostrate herb, sparingly branched growing up to 1.2 meters in length, and rooting at the lower nodes. The leaves are lanceolate, 4-10 mm long, and pointed at both ends. The flowers are very numerous, bright blue in colour, 8-10 mm across and borne in racemes. The sepals are hairy, green, oblong-linear, and about 5 mm long. The capsule is ovoid, surrounded by the persistent sepals, and contains numerous, minute, oblong seeds.

Ecology: It is commonly found in wet places like pond margins, stream sides, open forests, swampy or inundated soil and distributed throughout India.

Use: Leaves and young shoots are roasted and taken as food.

Hydrolea zeylanica (L.) Vahl
Vernacular names
- Languliya (O)
- Ceylon Hydrolea (E)
- Koilary (H)

50

Hygrophila auriculata (Schumach.) Heine

Hindi Name: Talimakhana

English Name: Marsh Barbel

Family: Acanthaceae

Flowering: October-December

Fruiting: January-February

Botanical description: The plant is a stout aquatic perennial herb growing up to 1-2 m high. The stems are erect, unbranched and hairy near swollen nodes. The leaves are sessile, linear- oblong, stalkless, 10-15 cm long, occur in whorls of 6 at each node on the stem. Straight, yellow, 4 cm long spines are present in the axil of each leaf. Flowers are purple in colour, occur in 4 pairs at each node and open in opposite pair.

Ecology: It is commonly found along water courses or wet places and distributed throughout India.

Nutritional value/Use: The plant is rich in protein and minerals. Leaves are roasted with oil and then eaten

Hygrophila auriculata (Schumach.) Heine
Vernacular names
- Khikri sag/Koelekha (O)
- Marsh Barbel (E)
- Talimakhana (H)

51

Ipomoea aquatica Forssk.

Hindi Name: Kalam Sag

English Name: Water Bindweed

Family: Convolvulaceae

Flowering: Most part of the year

Fruiting: Most part of the year

Botanical description: The plant is a semi-aquatic tropical herbaceous plant. The stem of the plant is thick, fistular or spongy, rooting at the nodes trailing on moist soil or mud, or floating on water. Leaves are variable in shape and size,such as ovate-oblong, ovate- triangular, lanceolate, elliptic-oblong.The flowers are trumpet-shaped, 3-5 cm diameter, usually white in colour with a purple center.Capsule ovoid, glabrous and 2-4 seeded.

Ecology: It is a common weed in and on the margins of ponds, tanks and swampy areas, naturalized throughout India. The plant is also cultivated presently because of its taste and market value.

Use: The leaves and tender shoots are collected, fried with oil and eaten

Nutritional Value: The plant is rich in vitamins and minerals. The nutritional value of the plant is given below. It is a rare source of manganese.100g of the plant contains about:

Carbohydrate 3.14g

Protein 2.6g

Fat 0.2g

Vitamin C (55mg), A (315μg), B_1 (0.03mg), B_2 (0.1mg), B_3 (0.9mg)

Mineral Ca (77mg), Mg (71mg), Fe (1.67mg), Mn (0.16mg)

Economic value: The plant is used as cattle feed

Ipomoea aquatica Forssk.
Common name
- Kalama saga (O)
- Water bindweed (E)
- Swampy Morning Glory (H)

52

Lagenaria siceraria (Molina) Standl.

Hindi Name: Lauki

English Name: Bottle Gourd

Family: Cucurbitaceae

Flowering: July-October

Fruiting: November-January

Botanical description: The plant is a vigorous, annual, climbing herb. It grows fast and may begin to flower only two months after seeding. The vine is branched and climbs by means of tendrils along the stem. The foliage is covered with soft hairs and has a foul musky odor when crushed. The leaves of the plant are up to 15 inches wide, circular in overall shape, with smooth margins, a few broad lobes, or with undulate margins. Leaves have a velvety texture because of the fine hairs, especially on the undersurface. The flowers are white in colour and borne singly on the axils of the leaves, the males on long peduncles and the females on short peduncles. Fruits are of various shapes like bottle or dumb bell-shaped, rounded or elongated.

Ecology: It is widely cultivated throughout India and in tropical and warmer regions of the world. The plant loves to grow in sandy soils.

Use: Leaves and young stems are cooked with other vegetables then eaten

Nutritional Value: The plant is rich in minerals and vitamins. The nutritional value of the plant is presented below. 100g of the plant contains about:

Carbohydrate 3.39g

Protein 2.56g

Fat 0.002g

Vitamin A-4.8μg, B_1-0.022mg

C-10.1mg

Mineral Ca-26mg, Fe- 0.20mg, P-13mg, Mg- 11mg

Economic value: The fruit of the plant is having nutritional importance and used as vegetable. The fruit is also used for making musical instrument.

Lagenaria siceraria (Molina) Standl.
Vernacular names
- Laoo (O)
- Bottle Gourd (E)
- Lauki (H)

53

Lepidium sativum L.

Hindi Name: Himba Saga

English Name: Garden Cress

Family: Brassicaceae

Flowering: June-August

Fruiting: August-October

Botanical description: The plant is an annual herb that reaches at a height upto 2 ft, with many branches on the upper part. The leaves are entire, the upper ones are stalkless whereas the lower ones are stalked. Flowers are white to pinkish in color, small, 2 mm across, borne in long racemes. The fruits are small pods, obovate, with two seeds per pod. The seeds are brownish red in colour and become slimy when soaked in water. Garden cress is native to mediterranean region, introduced and naturalised, cultivated throughout the world. Within India, it is cultivated as a salad plant and found as an escape.

Ecology: It is widely cultivated as vegetable, occasionally found as a weed in cultivated land in the tropical countries including India.

Use: Leaves and young shoots are collected, roasted and then eaten

Nutritional Value: The plant is rich in vitamins and minerals. The plant is a rich source of Magnesium, Iron and Calcium. The nutritional value of the plant is presented below. 100g of the plant contains about:

Carbohydrate 5.5g

Protein 2.6g

Fat 0.7g

Vitamin: C(69mg), E(0.7mg), A(346μg), B_1(0.01mg), B_2(0.26mg), B_3(1mg), B_6(0.274mg)

Mineral: Ca(81mg), Fe(1.3mg), Mg(38mg), Mn(0.553mg), P(73mg)

Lepidium sativum L.
Vernacular names
- Himba saga (O)
- Garden cress (E)

54

Leucas aspera (Willd.) Link

Hindi Name: Chhota halkusa

English Name: Garden Grass

Family: Lamiaceae

Flowering: July-October

Fruiting: October-January

Botanical description: The plant is an erect and diffusely branched annual herb, growing up to 20-60 cm in height. Leaves are linear or oblong, 2.5 to 7.5 cm long with blunt tips and scalloped margins. Whorls are large, terminal and axillary, about 2.5 cm in diameter and crowded with white bell shaped flowers. Calyx is variable, with an upper lip and short, triangular teeth. Nutlets are 2.5mm in size, smooth and brown.

Ecology: It is a common weed seen growing in waysides, waste places, cultivated lands. The plant also loves to grow in open dry sandy soils and distributed throughout India.

Use: Leaves are eaten after frying or roasting

Nutritional Value: The plant is rich in vitamin C, minerals and proteins.

Economic value: The plant is used as medicine for the treatment of wounds, cough and cold. Cattle loves to eat on this plant.

Leucas aspera (Willd.) Link
Vernacular names
- Gaiso (O)
- Garden grass (E)
- Chhota halkusa (H)

55

Limnophila indica (L.) Druce

Hindi Name: Kuttra

English Name: Indian Marshweed

Family: Plantaginaceae

Flowering: September-November

Fruiting: December-February

Botanical description: The plant is a perennial, small marsh herb. Stems are much branched, hairless. The aerial stems are up to 15 cm height, simple or branched, hairless, with stalked or stalkless glands. The submerged leaves are 1.5-2.5 cm in size, pinnately cut into segments which are flattened to capillary. The aerial leaves are usually whorled, pinnately lobed, 0.4-2 cm, sometimes few opposite and toothed. Flowers arise singly in the axils of aerial leaves. Flower stalk is slender, and 2-10 mm in length. Petals are botched with dark purple or reddish spots. The capsule is dark brown, compressed with ellipsoid to round in shape.

Ecology: It is common along borders of ponds, flooded paddy fields, wet low lands etc. The plant can survive in water as well as on land.

Use: Leaves and young shoots are roasted and taken as food. The plant is rich is minerals and protein.

Limnophila indica (L.) Druce
Vernacular names
- Keralata (O)
- Indian Marshweed (E)
- Kuttra (H)

56

Marsilea minuta L.

Hindi Name: Sunsuni

English Name: Dwarf Water Clove

Family: Marsileaceae (Pteridophyta)

Flowering: November-January

Fruiting: January-March

Botanical description: The plant is a branched herbaceous plant with creeping rhizome and usually 3-15 cm or more in length. Leaves are composed of 4 obdeltoid leaflets. Leaflets are cruciform, obovate, variable in size depending on the ecological conditions, usually glabrous with rounded outer margin. Pedicels are 2-6, basal and slightly connate or free. Sporocarps are seen crowded or in groups of 2–3, arising at the base of the petiole.

Ecology: It is widely found in wet places like shallow pools, edges of rivers, rocks underneath waterfalls, seasonally dried up river beds, rice fields etc. and distributed throughout India.

Use: Tender leaves and young shoots are roasted and taken as food

Nutritional Value: The plant is a rich source of iron, magnesium and vitamin C. The nutritional value of the plant is presented below. 100g of the plant contains about:

Carbohydrate 2.3g

Protein 1.53g

Fat 0.002mg

Vitamin: B1-0.016mg, C- 2.85mg

Mineral: Ca (110mg), Fe (5.07mg), Mg (136mg),

Marsilea minuta L.
Vernacular names
- Sunsunia (O)
- Dwarf Water Clover (E)

57

Mentha spicata L.

Hindi Name: Pudina

English Name: Spearmint

Family: Lamiaceae

Flowering: August-September

Fruiting: September-October

Botanical dcescription: The plant is a glabrous perennial herb. Stems are erect, about 1 ft tall, usually less, purplish, hairless. Leaves are stalkless, sessile, lanceolate or ovate, acute with glabrous above and glandular below. Flowers are lilac, loose cylindrical, slender with interrupted spikes.

Ecology: It is widely cultivated in India and often planted in kitchen gardens. The plant loves to grow in sandy soils.

Use: Young leave paste added with salt and chilly is used as chantey

Nutritional Value: The plant is a rich source of vitamin C and iron. The detail nutritional value of the plant is described below. 100g of the plant carries approximately:

Carbohydrate 8g

Protein 3.3g

Fat 0.7g

Vitamin A-2.43mg, C-13.3mg, B1-0.078mg B2-0.17mg

Mineral Ca-199mg, Fe-11.87mg, Mg-63mg

Economic value: The plant has got medicinal value and is used for treatment of stomach disorders.

Mentha spicata L.
Vernacular names
- Podina (O)
- Spearmint (E)
- Pudina (H)

58

Mitracarpus hirtus (L.) DC.

Hindi Name: Lauki

English Name: Tropical girdlepod

Family: Rubiaceae

Flowering: September-December

Fruiting: December-March

Botanical description: The plant is erect or diffused, pubescent herb, growing up to 40 cm in height. The leaves are sessile, ovate-lanceolate, acute, pubescent and rounded at the base. Flowers are small, white in colour and verticillate in axillary fascicles with small calyx. Capsule is globose, less than 1 mm across with 2 minute seeds and having furrowed on one side.

Ecology: It is a common weed in waste or cultivated grounds, and sometimes found in in pastures, on open banks, or in thickets and distributed throughout India.

Use: Leaves are boiled with water and then eaten.

Economic value: The plants having medicinal importance, are especially eaten to cure dysentery and diarrhoea.

Mitracarpus hirtus (L.) DC.
Vernacular names
- Ganthia (O)
- Tropical girdlepod (E)
- Lauki (H)

59

Momordica charantia L

Hindi Name: Karela

English Name: Bitter Gourd

Family: Cucurbitaceae

Flowering: June-October

Fruiting: October-February

Botanical description: Bitter gourd is a herbaceous, tendril-bearing, slender climber, growing up to 5 m. It bears simple, alternate leaves 4-12 cm across, with 3-7 deeply separated lobes. Each plant bears separate yellow male and female flowers, about 2-3 cm in diameter. Male flowers are more numerous, having a yellow center and conical base, while female flowers have a green center and a small bump at the base. The fruits are oblong shaped and tapering at both ends, 5-20 cm long and turns yellow when ripe. It is hollow in cross-section, with a relatively thin layer of flesh surrounding a central seed cavity filled with large flat seeds and pith. Seeds and pith appear white in unripe fruits.

Ecology: It is common in hedges and thickets, along creeks and streams and also commonly cultivated all over India.

Use: Leaves are fried with other vegetables and then eaten. Juice of the leaves are also taken to stop worm infection

Nutritional Value: The leaf of the plant is a good source of vitamin A. The plant is also rich in minerals and other vitamins. The nutritional value of the plant is presented below. 100g of the plant contains about:

Carbohydrate 3.70g

Protein 1.00g

Fat 0.17g

Vitamin A-290mg, C- 84mg, B1-0.040mg

Mineral Ca-19mg, Fe-0.43mg, Mg-17mg

Economic value: The plant having medicinal importance, especially eaten to cure dysentery and skin disorder.

Momordica charantia L
Vernacular names
- Kalara (O)
- Bitter gourd (E)
- Karela (H)

60

Moringa oleifera Lam.

Hindi Name: Sanjena

English Name: Drum Stick

Family: Moraginaceae

Flowering: January-March

Fruiting: April-June

Botanical description: The plant is a small deciduous tree, having corky gray bark, branching and fernlike leaves and growing up to 10 metre in height. The leaves are pinnately compound, 25-50 cm long with thickened and articulated at the base. The flowers of the plant are scented and white in colour with lanceolate petals. Capsule is pendulous, long bean like seed pods and usually 22-45 cm long.

Ecology: It grows in the soils that are well drained and the water table remains fairly high all the year round and also commonly cultivated as vegetable in the tropical countries including India. Often cultivated in dry and waste lands.

Use: Leaves are eaten after frying or roasting with oil and also used to prepare curry. The leaves are extensively used as a vegetable in many parts of the world. The fruits are also used to prepare curry, sambar etc. popular in eastern India.

Nutritional Value: The plant is rich a rich source of iron and magnesium. The nutritional value of the plant is presented below. 100g of the plant carries:

Carbohydrate 8.28g

Protein 9.40g

Fat 1.40g

Vitamin: C(51.7mg), B_6(1.200mg), B_5(0.125mg), B_3(2.220mg), B_2(0.660mg)

Mineral: Ca(185mg), Fe(4.00mg), Mg(147mg), Mn(0.36mg), P(337mg)

Economic value: The plant is having medicinal importance, bark juice is used as medicine for snake bite and nasal congestion. The juice of the plant leaves helps in reducing blood pressure.

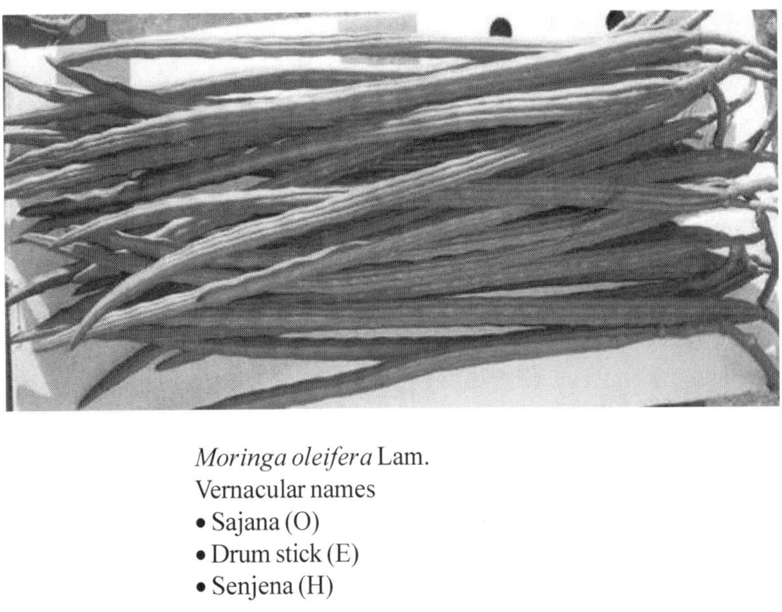

Moringa oleifera Lam.
Vernacular names
- Sajana (O)
- Drum stick (E)
- Senjena (H)

61

Murraya koenigii L. Spreng

Hindi Name: Barsanga

English Name: Curry Tree Leaf

Family: Rutaceae

Flowering: February-April

Fruiting: May-September

Botanical description: The plant is a small tree, growing up to 4-6 m tall, with a trunk up to 40 cm diameter. The leaves are opposite or alternate, lanceolate or ovate, strongly scented with 11-21 leaflets. The flowers are small white, and fragrant. The berries are small, ovoid or ellipsoid, black in colour, shiny and edible, but their seeds are poisonous.

Ecology: It is cultivated and also grows wild. The plant usually grows in dry low land areas and is distributed throughout India.

Use: Leaves are used as spice in curry and it gives a specific scent. Also young leaves are used to prepare chatny.

Nutritional Value: The plant is rich in minerals and vitamins, especially calcium and magnesium. The nutritional value of the plant is presented below. 100g of the plant contains about:

Carbohydrate 18.70g

Protein 6.10g

Fat 1.00g

Vitamin A-756μg, B_1-0.210mg, C-4mg

Mineral Ca-830mg, P-57mg, Fe-0.93mg, Mg-44mg

Economic value: The plant having medicinal value is used for stomach disorder and dysentery. It has got high antioxidant property.

Murraya koenigii L. Spreng
Vernacular names
- Bhursanga (O)
- Curry tree leaf (E)
- Barsanga (H)

62

Oroxylum indicum (L.) Kurz

Hindi Name: Bhut-vriksha

English Name: Midnight horror

Family: Bignoniaceae

Flowering: July-August

Fruiting: December-March

Botanical description: The plant is a medium sized deciduous tree, growing up to 8-12 m tall. The bark is grayish brown with corky lenticels. The leaves are huge, 0.5-1.5 m long, 2-3 pinnate with leaflets of 12 cm long and 8 cm broad. The flowers are reddish- purple outside and pale, -yellowish within, numerous, in large erect racemes. The fruits are flat capsules, 0.33-1 meter long and 5-10 cm broad, sword-shaped. The seeds numerous, flat and winged all around, except at the base.

Ecology: It is occasional and seen in in mixed deciduous forests, especially in the hills, also in waysides and distributed throughout India.

Use: Leaves are eaten after frying or roasting.

Economic value: The plant having medicinal value is used for Skin diseases.

Oroxylum indicum (L.) Kurz
Vernacular names
- Phapni saga (O)
- Midnight horror (E)
- Bhut-vriksha (H)

63

Oxalis corniculata L.

Hindi Name: Amurul

English Name: Indian Sorrel

Family: Oxalidaceae

Flowering: June-August

Fruiting: September-October

Botanical description: The plant is a creeping herb, branching from the base and often rooted at the nodes, the upper portion is ascending or weakly erect, smooth or hairy. The leaves are arranged alternately along the stems. A single long stalk arises from the axils of the leaf, from which extend three flower stalks, each with a single flower. The flowers are 7-11 mm wide and have 5 yellow petals. The fruit is a capsule, 1-1.5 cm long, cylindric, pointed apically, and 5-ridged in cross section.

Ecology: It is a fairly common weed found in gardens, elsewhere such as arable land and waste places and distributed throughout India.

Use: Leaves are cooked with mustard oil, chilly, salt with potato as delicious curry to eat.

Nutritional Value: The plant is a good source of vitamin C and also rich in protein and minerals. The nutritional value of the plant is presented below. 100g of the plant contains about:

Carbohydrate 24.67g

Protein 22.28g

Fat 2.3g

Vitamin 19.6g

Mineral Ca (2.5g), Mg (0.25g), P(2.17mg), Na (1.12mg)

Economic value: The plant is used as medicine for treatment of hookworms in children.

Oxalis corniculata L.
Vernacular names
- Amliti saga (O)
- Indian sorrel (E)
- Amurul (H)

64

Paederia foetida L.

Hindi Name: Gandhali

English Name: Skunk vine

Family: Rubiaceae

Flowering: August-September

Fruiting: September-October

Botanical description: The plant is a perennial twining climber, arising from a woody rootstock. Its stems grow up to 7 m or more, climbing, or prostrate and rooting at the nodes. Leaves are oval to linear-lanceolate, oppositely arranged, 2-11 cm long, hairy or smooth, often lobed at base with prominent stipules. Leaf stalks are mostly up to 6 cm long. Leaves and stems have a atinking smell, especially when crushed. Flowers are small, purple, broad or long, curving clusters at the end of branches or in leaf axils. Fruit is a shiny brown, nearly round capsule, up to 0.7 cm across, with 2 black, rounded seeds.

Ecology: the plant is seen growing on hillsides, in forests, along forest edges, along streamsides, and twining on trees and distributed throughout India.

Use: Leaves are eaten after frying or roasting

Economic value: The leaves of the plants also having medicinal importance, used as medicine against diarrhea and dysentery.

Paederia foetida L.
Vernacular names
- Pasaruni (O)
- Skunk vine (E)
- Gandhali (H)

65

Phyla nodiflora (L.) Greene

Hindi Name: Jal buti

English Name: Frog Fruit

Family: Verbenaceae

Flowering: August-September

Fruiting: September-October

Botanical description: The plant is a prostrate, flowering and broadleaf herb. The leaves are opposite, obovate, serrate above middle, obtuse or rounded and thick. The flowers are small, in axillary peduncled and consists of a purple-coloured centre encircled by small white-to-pink flowers. Fruits are 2 mm long, enclosing two with 1-celled pyrenes.

Ecology: It is frequently seen in open moist places near water courses and distributed throughout India.

Use/Economic value: Leaves are collected, then fried with oil and taken after cooking

Phyla nodiflora (L.) Greene
Vernacular names
- Langli saga/Gosingi (O)
- Frog Fruit (E)
- Jal buti (H)

66

Pimpinella heyneana (Wall. ex DC.) Kurz

Hindi Name: Charmal

English Name: Heyne burnet

Family: Fabaceae

Flowering: October-November

Fruiting: November-December

Botanical description: The plant is an annual, erect branched, glabrous herb growing upto 30-90 cm in height. Leaves are compound, leafelets are lanceolate, often doubly serrate, acuminate and sparsely puberulous both side with acute to rounded base. The flowers are 9-16 in number and arranged in umbellule, 1.5 mm across, with a pedicels 5 mm long. Fruit is ovate, slightly compressed laterally, glabrous with prominent ribs.

Ecology: It is commonly seen growing in hill forests and distributed throughout India.

Use: Leaves along with flowers and seeds are collected, fried then eaten. The plant is having a special fragrance for which it is used as a spice in cooking.

Economic value: The plant also sold in weekly markets of tribal regions for its demand because of food value.

Pimpinella heyneana (Wall. ex DC.) Kurz
Vernacular name
• Charmali (O)
• Heyne burnet (E)

67

Pisum sativum L.

Hindi Name: Matar

English Name: Pea

Family: Fabaceae

Flowering: November-December

Fruiting: January-February

Botanical description: The plant is smooth, glaucous, annual herbaceous plant with foloaceous ovate stipules, irregularly toothed at the base. The leaflets are 2-3 pairs, opposite leaf-rachis ending in a branched tendril. The colour of the flower is white, peduncle is 1 and few flowered. Pods are broad with rounded, green or yellowish seeds.

Ecology: It is widely cultivated as vegetable throughout India. The plant usually grows on sandy and loamy soil, often grows on rocky soil.

Use: Leaves and young shoots are collected, fired and eaten.

Nutritional Value: The plant is rich in vitamins and minerals, especially magnesium and zinc. The nutritional value of the plant is presented below. 100g of the plant contains about:

Carbohydrate 10.45g

Protein 6.42g

Fat 0.40g

Vitamin A-256 μg, B1-0.132mg, C-40mg, E-0.13mg

Mineral Ca-25mg, Fe-1.47mg, Mg-33mg, Zn-1.24mg

Economic value: The seeds of the plant are commonly used as vegetables.

Pisum sativum L.
Vernacular names
- Matar saga (O)
- Pea (E)
- Matar (H)

68

Plectranthus mollis (Aiton) Spreng.

Hindi Name: Lal Aghada

English Name: Soft-Stem Mintleaf

Family: Lamiaceae

Flowering: September-October

Fruiting: November-December

General Characters: The plant is a small, erect, fleshy, annual herb, growing up to 30-50 cm high. The leaves are oppositely arranged, broadly ovate, 5-12 cm long, pointed, heart- shaped at base, with a toothed margin. Flowers are pale blue in colour, borne in branched racemes, 7-20 cm long, at the end of branches. Fruits are round, 2-3 mm, gery brown, with sparse small dark warts.

Ecology: It is frequently seen in forests, anthropogenic habitats, wastelands under shade, especially in the hills and is distributed throughout India.

Use: Leaves are fried with oil and chilly then eaten. Seeds are also used to prepared sweets.

Economic value: The plant is having medicinal value. Leaf juice is used against headache and fever.

Plectranthus mollis (Aiton) Spreng.
Vernacular names
• Sakra/Gondi (O)
• Soft-Stem Mintleaf (E)
• Lal Aghada (H)

69

Polygonum plebeium R. Br.

Hindi Name: Machechi

English Name: Alpine knot weed

Family: Polygonaceae

Flowering: February-April

Fruiting: April-June

Botanical description: The plant is a prostrate and densely branched, annual herb. Stems are branched from base, branches numerous. The leaves are linear, oblong 2.5-18 mm long, acute or obtuse and often with revolute margins. Flowers are small, pink with green midrib, subsessile and sunk among the stipules or exsert. The nuts are 1.0-1.75 mm long, circular to ovate, shining, and black.

Ecology: It is a fairly common weed in moist places, cultivated ground etc and distributed throughout India.

Use: Leaves and young shoot are roasted with mustard oil and then eaten

Nutritional Value (100gm): The plant is rich in protein and minerals, especially sulphur and magnesium. The nutritional value of the plant is presented below. 100g of plant contains about:

Carbohydrate 12.5g

Protein 14.9g

Fat 1.24g

Vitamin C-71.79mg

Mineral Ca-27.44mg, Mg-17.73mg, P-10.13mg, S-11.37mg

Polygaonum plebeium R.Br.
Vernacular names
- Muthi saga (O)
- Alpine knot weed (E)
- Machechi (H)

70

Portulaca oleracea L.

Hindi Name: Khursa

English Name: Common Purslane

Family: Portulacaceae

Flowering: February-April

Fruiting: April-June

Botanical description: The plant is prostrate or erect herb growing upto10-30 cm in height. Leaves are alternate or subopposite, margins entire, obtuse or truncate apex, clustered at stem joints and ends. The flowers are yellow in colour, sessile to 1 cm across, open singly at the center of the leaf cluster for only a few hours on sunny mornings. The seeds are brown in colour, 0.8 mm with minutely tubercled.

Ecology: It is a very common weed seen in cultivated lands, fields, wastelands, waysides, by the sea, near water courses etc. and widely distributed throughout India.

Use: Tender leaves and shoots are collected, fried with other vegetable then eaten.

Nutritional Value: The plant is rich in vitamins and minerals. The nutritional value of the plant is given below. 100g of the plant carries about:

Carbohydrate 3.4g

Protein 1.30g

Fat 0.1g

Vitamin B1-0.480mg, B2-0.112mg, B3-0.480mg, C-21mg

Mineral P-44mg, Ca-65mg, Fe-1.99mg

Economic value: Because of nutritional value, the plant has got market value.

Portulaca oleracea L.
Vernacular names
- Balubalua saga (O)
- Common purslane (E)
- Khursa/Kulfa (H)

71

Raphanus sativus L.

Hindi Name: Muli

English Name: Radish

Family: Brassicaceae

Flowering: December-January

Fruiting: January-February

Botanical description: The plant is an edible root vegetable growing upto 20-40cm in height. leaves usually are medium green and lobed and have a rough texture, lower leaves are lyrate, upper ones is lobed. The flowers are cross-shaped with four petals, usually about an inch across and white, pink or purple in colour. The seed pods are sickle shaped, 2-12" long, first green, then turning brown when mature. They are full of little round seeds.

Ecology: It is widely cultivated as vegetable in India.

Use: Leaves are fried along with other vegetables and eaten. The modified root is used as vegetable and in salads.

Nutritional Value: The plant is rich in vitamin C, K besides potassium, magnesium and sodium. 100g of the plant contains about:

Carbohydrate 10.6g

Protein 4.4g

Fat 0.2g

Vitamin A-330μg, B1-0.18mg, C-55mg, E-3.5mg, K-540 μg

Mineral P-400mg, Na-42mg, Ca-210mg, Mg-22mg

Economic value: The edible root is very nutritive and commonly used as vegetable and also used as medicine for digestive disorder. The leaves are eaten by cattles as well.

Raphanus sativus L.
Vernacular names
- Mulasaga (O)
- Radish (E)
- Muli (H)

72

Rivea hypocrateriformis Choisy

Hindi Name: Thor-Ki-Bel

English Name: Midnapore Creerper

Family: Convolvulaceae

Flowering: October-November

Fruiting: November-December

Botanical description: The plant is a large climber with silky branches. The leaves are broadly ovate, 5-8.7 cm in diameter with obtuse apex. Flowers are large, yellowish-white in colour, fragrant, 6 cm long, 5 cm across on 1-3 flowered with long tubes. Fruit is globose, 1.2 cm in diameter, enclosed by enlarged and woody calyx.

Ecology: It is usually found in forests and hilly areas and distributed throughout India.

Use: Leaves are collected, fried and eaten

Nutritional Value: The plant is rich in carbohydrate and protein. The nutritional value of the plant is presented below. 100g of the plant contains about:

Carbohydrate 57.63g

Protein 19.27g

Fat 2.66g

Mineral Ca-0.99g, Mg-0.34g, P-0.32g

Rivea hypocrateriformis Choisy
Vernacular names

- Kindu saga (O)
- Midnapore Creeper (E)
- Thor-Ki-bel (H)

73

Rungia pectinata (L.) Nees

Hindi Name: Sut

English Name: Comb Rungia

Family: Acanthaceae

Flowering: November-December

Fruiting: January-February

Botanical description: The plant is a diffused or erect herb with cylindrical stem and growing up to 10-20 cm in height. The leaves are oppositely arranged, nealy glabrous, oblong-lanceolate with sharp-tipped. Flowers are tiny, violet-blue in colour and are about 3-4 mm long, and occur in spikes at the end of branches, or in leaf axils. The capsule in 2 mm in size and compressed.

Ecology: The plant is a fairly common as weed, abundantly found on rocky surfaces and distributed throughout India.

Use: Leaves are collected, fried with oil then eaten. The plant is rich in nutrient.

Rungia pectinata (L.) Nees
Vernacular names
- Hasa ada/Mati sag (O)
- Comb Rungia (E)
- Sut (H)

74

Senna tora (L) Roxb

Hindi Name: Panwar

English Name: Sick Senna

Family: Caesalpiniaceae

Flowering: September-November

Fruiting: November-December

Botanical description: The plant is a small erect hairlesss shrub, growing about 1 m tall. The leaves are compound, arranged spirally and usually have three pairs of symmetrically egg- shaped, leaflets up to 2 inches long. One to three yellow flowers appear on short axillary stems. The linear pods grow upto 8 inches long, curve downward and contain many shiny, and angular seeds.

Ecology: It occurs abundantly in open pastures, and is very common on roadsides, wastelands and distributed throughout India.

Use: Tender leaves and shoots are cooked and then taken as food

Economic value: The plant is used as medicine against skin disease, worm infection, asthma and piles

Senna tora (L) Roxb
Vernacular names

- Chakunda saga (O)
- Sickle Senna (E)
- Panwar (H)

75

Solanum americanum Mill.

Hindi Name: Mokoi

English Name: Black Nightshade

Family: Solanaceae

Flowering: July-September

Fruiting: October-December

Botanical description: The plant is an erect, branched herbaceous weed that grows up to 60cm tall. The leaves are alternate, ovate or oblong, deep green in colour with an indented margin and acuminate at the tip. Flowers are small, white in colour with 5-1.5 mm across. The berries are globose in shape, green at early stage and turn to orange or black when ripened.

Ecology: It is a fairly common weed in, thickets, often on cultivated or waste ground, waste places, roadsides and distributed throughout India.

Use: Leaves are collected, fried, then eaten

Economic value: The plant is having medicinal importance and used against fever and tooth ach.

Solanum tuberosum L.
Common name:
- Alu saga (O)
- Potatao (E)
- Alu (H)

76

Solanum tuberosum L.

Hindi Name: Alu

English Name: Potato

Family: Solanaceae

Flowering: November-December

Fruiting: January-March

Botanical description: The potato plant is a leafy, tufted herb with tuberous underground stems growing up to 2 ft tall and spreading out a little more. The leaves are compound, long and the leaflets are 7-15 in number, with about 3 inch in long. The tubers are not roots, but modified stems or rhizomes, and the "eyes" are really leaf buds. The flowers are star-shaped, white or blue in colour and about an inch across, and borne in clusters. The fruits are like small green tomatoes, about an inch in diameter, and contain several hundred seeds.

Ecology: It is a cultivated throughout India and loves to grow in sandy and loamy soil.

Use: Leaves are collected, fried separately or with other vegetables, then eaten. Tubers are boiled or fried and eaten.

Economic value: The plant is having medicinal importance and used against Fever and tooth pain.

Solanum americanum Mill.
Common name
- Nunununia (O)
- Black Nightshade (E)
- Mokoi (H)

77

Spinacia oleracea L.

Hindi Name: Paalak

English Name: Spinach

Family: Chenopodiaceae

Flowering: November-December

Fruiting: January-March

Botanical description: The leaves of the plant are alternate, lower most and middle one is long petioled, ovate-oblong, obtuse or acute, entire or dentate and higher one is small. Flowers are dioecious. Male flowers in arranged in spicate clusters combined into a leafy terminal panicle and female flower in axillary clusters. Ovary enclosed by tepal like bracts that developed into a globose or pyramidal, 2-4 spined structure in fruiting.

Ecology: It is a annual herb, commonly cultivated and often self shown throughout India. The plant usually grows on sandy and loamy soils.

Use: Leaves are fried with vegetables and also cooked with dal and potato to make curry.

Nutritional Value: The plant is rich in vitamins (C, B complex) and minerals (Iron, calcium etc.). The nutritional value of the plant is presented below. 100g of the plant contains about:

Carbohydrate 3.6g

Protein 2.9g

Fat 0.4g

Vitamin C(28mg), E (2mg), A(469μg), B_1 (0.078mg), B_2(0.189mg), B_3(0.724mg), B3(0.159mg)

Mineral Ca(99mg), Fe(2.71mg), Mg(79mg),

Mn(0.897mg), P(49mg), K(558mg)

Economic value: Commonly cultivated as vegetable that has good market value.

Spinacia oleracea L.
Common name
- Palanga (O)
- Spinach (E)
- Palak (H)

78

Talinum portulacifolium (Forssk.) Asch. ex Schweinf.

Hindi Name: Ajwain

English Name: Ceylon spinach

Family: Talinaceae

Flowering: September-October

Fruiting: November-December

Botanical description: The plan is an erect, stout, fleshy herb, growing up to 0.5-1 m in height. Leaves are elliptic to obovate, 5-15 cm long, 1.5-5 cm wide, tip pointed to tapering. The flowers are borne in 2-5-branched, raceme like clusters, 3-20 cm long with a stalk about 8-11 mm long. Petals are pink, obovate, up to 1 cm long. Capsules are pale yellow, ellipsoid, 6-7 mm long. Then plant flowers and bears fruits during September-December.

Ecology: It is fairly common weed in thickets, often on cultivated or waste ground, waste places, roadsides and distributed throughout India.

Use: Leaves are collected fried with other vegetable then eaten

Talinum portulacifolium (Forssk.) Asch. ex Schweinf.
Vernacular names
- Mati poi (O)
- Ceylon spinach (E)

79

Trachyspermum ammi (L.) Sprague

Hindi Name: Ajwain

English Name: Indian Dill

Family: Apiaceae

Flowering: November-January

Fruiting: February-April

Botanical description: The plant is an erect, freely branching annual herb with finely dissected, lacy, blue-green leaves. The leaves are about 1 ft long and divided pinnately three or four times into thread like segments each about 1 in long. The dill plant grows about 3-5 ft tall. Occasionally it becomes top-heavy and falls down. The flowers are white and borne in large, rounded, compound umbels, like all carrot family flowers. The umbels are borne on stiff, hollow stems. The whole inflorescence can be 25 cm across. The fruit is a flattened pod about 3 mm long. All parts of the dill plant are strongly aromatic.

Ecology: It is widely cultivated or often found as an escape in the tropical countries including India.

Use: Leaves and young shoots are collected, fried and then eaten. It is highly scented.

Nutritional Value: The plant is rich in minerals and protein especially calcium, potassium and magnesium. The nutritional value of the plant is presented below. 100g of the plant contains about:

Carbohydrate 55.8g

Protein 20g

Fat 4.4g

Vitamin 14 g

Mineral Ca: 1784mg;P: 543mg Fe: 48.8mg

Mg: 451mg

Economic value: The plant is used as medicines for stomach disorder and dyscentry.

T*rachyspermum ammi* (L.) Sprague
Vernacular names

- Juani (O)
- Indian dill (E)
- Ajwain (H)

80

Trianthema portulacastrum L.

Hindi Name: Kasturi Methi

English Name: Wild Trefoil

Family: Aizoaceae

Flowering: July-September

Fruiting: October-December

Botanical description: The plant is a succulent herb with prostrate forked branches, growing up to 20-45 cm long. Stems are prostrate or rising, somewhat succulent, up to 50 cm long or more, smooth or sparsely velvety. Leaves are flat, elliptic to obovate, 1-2 cm long, 0.4-2 cm wide, margins entire, tip blunt with rounded base. Leaf stalks are 0.5-2.5 cm long, expanded into a sheath joined with leaf base to form a cup. Flowers are pinkish or white in colour, borne solitary, stalkless, largely hidden in leaf axils. Capsule with a lower scarious or membranous portion and an upper brown. Seeds are black in colour with muricate concentric lines.

Ecology: It is a common weed, occurs in wastelands, roadsides, lawns, gardens, cultivated crops, and in paddy fields if the water supply is low.

Use: Leaves and young shoots are collected, fried separately or with other vegetables to prepare curry, then eaten.

Nutritional Value: The plant is rich in mineral, carbohydrate and vitamin B2. the nutritional value of the plant is presented below. 100g of the plant contains approximately:

Carbohydrate 30.2g

Protein 24.6g

Fat 1.69g

Vitamin: A-896 mg, B2-2.02g

Mineral: Mg-0.654mg, Fe-86.3mg, Na-44.8 mg

Economic value: The leaves are also used as medicine for external wound injury.

Trigonella portulacastrum L.
Vernacular names
- Firingi saga (O)
- Wild trefoil (E)
- Kasturi methi (H)

81

Trigonella corniculata L.

Hindi Name: Kasturi Methi

English Name: Wild Trefoil

Family: Fabaceae

Flowering: September-October

Fruiting: November-January

Botanical description: Thye plant is a small diffused herb, with erect to creeping habit. Leaflets are 1-4 cm long, 8-35 mm broad, obovate to oblong-wedge-shaped, tip rounded to notched. The flowers of the plant yellow in colour, bloom in 8-20-flowered clusters, with a stalks 3 mm long and sepals nearly equal, shorter than or as long as the tube. The fruit is 1-2 cm long, 2-3 mm broad, linear, sharp tipped, curved and 4-8-seeded.

Ecology: It is usually cultivated and also found in dry banks and grassy places and distributed throughout India.

Use: Leaves are fried along with other vegetables. Dried leaves are used in cooking as kastoori methi.

Trigonella corniculata L.
Vernacular names
• Firingi saga (O)
• Wild trefoil (E)
• Kasturi methi (H)

82

Trigonella foenum-graecum L.

Hindi Name: Methi

English Name: Fenugreek

Family: Fabaceae

Flowering: December-January

Fruiting: February-March

Botanical description: The plant is an erect, annual and strongly smelling herb growing up to 10-50 cm in height. The Leaves are trifoliate with leaflets 1-3 cm long, 5-15 mm broad, obovate to inverted-lanceo shaped. Flowers are cream or yellowish-white, in colour, 1-2 together in leaf axils and 1.2-1.8 cm long. Pods are oblong, straight or falcate with age, 7.5-10 cm long with a long beak and 20 seeded.

Ecology: It is commonly cultivated throughout India.

Use: Fresh leaves are collected and chutney is prepared or roasted and taken

Nutritional Value: The plant is rich in minerals and vitamins. The nutritional value of the plant is presented below. 100g of the plant contains:

Carbohydrate 12.6g

Protein 5.2g

Fat 0.6g

Vitamin C(18mg), A(60 IU), B1(0.332mg), B2(0.336mg), B3(1.640mg),

Mineral Ca(176mg), Co(1.110mg) Fe(33.53mg), Mg(191mg), Mn(1.228mg),

Economic value: The plant is having medicinal value used against digestive problems such as stomachache, constipation, inflammation of the stomach.

Trigonella foenum-graecum
Vernacular names
- Odia- Methi (O)
- Fenugreek (E)
- Methi (H)

83

Vicia sativa L.

Hindi Name: Choni

English Name: Black Pod Vetch

Family: Fabaceae

Flowering: December-January

Fruiting: February-March

Botanical description: The plant is a scrambling and climbing herb, with stems almost hairless, up to 3 ft tall. The leaves are compound with 5-6 pairs of opposite leaflets and 1 terminal tendril. Leaflets are narrow, linear, 2.5-3.0 cm long, 2-3 mm wide. Flowers are borne singly or in pairs on short stalks arising at the base of the leaves, blue to purple in colour, large, up to 2 cm long. Seed pods are mainly black, linear, 4-6 mm broad, containing 10-12 round or angulated small seeds. The plant flowers and bears fruits during December-March.

Ecology: The plant is a common weed of cultivation and usually found in field edges, waste places and also in hedges and grassy places.

Use: Leaves are roasted with oil and eaten

Vicia sativa L.
Vernacular names
- Anisaga/Choni (O)
- Black-Pod Vetch (E)

84

Aegle marmelos (L.) Correa

Hindi Name: Bel

English Name: Wood Apple

Family : Rutaceae

Flowering: March-April

Fruiting: (Next Year) April-June

Botanical Description: The plant is a medium sized tree with spines of 1-2 cm long. The leaf is composed of three leaflets. The leaflets are oval, elliptic with 2.5-6.5 cm petiole, terminal one is larger. Flowers of the tree are white or greenish-white. The fruits are ovoid, hard and about 5-7.5 cm in diameter. The flowers appear during March-April and the tree bears fruits towards October, ripening in April.

Ecology: It is common in scrub and open forests, also planted near temples and villages. The plant is usually distributed throughout the greater part of India.

Use: The fruit is having aromatic pulp which is eaten and carries a lots of nutritive value.

Nutritional Value: The plant is rich in carbohydrate, vitamins (C, B complex) and minerals (Iron, calcium). The nutritional value of the plant is presented below. 100g of the plant contains:

Carbohydrate 31.8g

Protein 1.8g

Fat 0.3g

Vitamin A (55mg) C(60mg), B2(1.19mg), B1(0.13mg), B3(1.1mg),

Mineral Ca(85mg), Fe(1mg), P(50mg)

Economic value: The leaves have medicinal value as well as religious importance. It is used for treatment of stomach disorders and for promoting digestion. It helps in lever functions It also used as holy leaf for different festivals and pujas, especially to worship Lord Shiva, the king of god.

Aegle marmelos L. Correa
Vernacular names
- Bela (O)
- Wood apple (E)
- Bel (H)

85

Artocarpus heterophyllus Lam.

Hindi Name: Kathal

English Name: Jack Fruit

Family : Moraceae

Flowering: December-January

Fruiting: June-July

Botanical Description: The plant is an evergreen tree that grows up to 20-30 meter tall and 80-200 cm in diameter with evergreen, alternate, glossy and leathery leaves reaching up to 9 inch long. Leaves are oval on mature trees. All parts of the plant contain sticky and white latex. Inflorescence is solitary, borne axillary on special lateral short leafy shoots arising from older branches. Fruits are large arising from the tree trunk and weighting about 10-50 kg. the exterior of the fruit is green or yellow when riped and composed of numerous hard, cone like points attached to a thick and rubbery, pale yellow or whitish wall.

Ecology: The plant found in wild and cultivated, seen in damp forests and waste places close to villages. It was introduced and has become naturalized in many parts of India.

Use: The fruits have a lot of nutritive value. It is consumed as raw vegetable and fruit after ripening. Unripe fruits are used as vegetable or making pickles.

Nutritional Value: The plant is rich in carbohydrate vitamins (C, B complex) and minerals (Iron, calcium). The nutritional value of the plant is presented below. 100g of the plant contains about:

Carbohydrate 23.5g

Protein 1.72g

Fat 0.64g

Vitamin C(13.7mg), E(0.3mg), A(110 IU), B

complex(0.055mg), B1(0.105mg), Mineral Ca(34mg), Fe(0.60mg), Mg(37mg),

Mn(0.197mg), P(36mg), Zn(0.42mg)

Economic value: The plant has high economical value because of its timber and fruit. The leaves are also used in various religious purposes and pujas.

Artocarpus heterophyllus Lam.
Vernacular names

- Panasa (O)
- Jack Fruits (E)
- Kathal (H)

86

Averrhoa carambola L.

Hindi Name: Karmal

English Name: Carambola

Family : Oxalidaceae

Flowering: July-September

Fruiting: October-December

Botanical Description: The plant is a small evergreen tropical tree that grows upto 25 feet in height. The compound leaves are soft, medium green, spirally arranged around the branches in an alternate fashion. The leaflets are sensitive to light and fold inward at night. The flowers are small and bell shaped, with five petals that have whitish edges. The fruit is fleshy, have a thin and waxy skin, five lobbed, ovate to ellipsoid with attractive yellow-orange in colour.

Ecology: It is commonly found open forests, waysides, wastelands etc. and often cultivated.

Use: The fruits are consumed as raw or cooked. The fruit is crisp and juicy with a waxy yellow skin when ripe. Ripe fruits are sweet in taste and consumed fresh, but sour fruits are used to prepare drinks, pickles or used as a substitute to tamarind. It is a good source of vitamin A, B and C.

Nutritional Value: The fruit is a good source of Vitamin C, E and Zinc.The nutritional value of the plant is described below. 100g of the plant (fruit) contains approximately:

Carbohydrate 6.73g

Protein 1.04g

Fat 0.33g

Vitamin B1-0.014mg, B2-0.016mg,

B3-0.367mg, C-34.4mg, E-0.15mg

Mineral Ca-3mg, Fe-0.08mg, P-12mg, Mg-10mg, Zn-0.12mg

Economic value: The fruit have got medicinal value, as a decoction of the fruit is used as a treatment for indigestion.

Averrhoa carambola L.
Vernacular name
- Karamanga (O)
- Carambola (E)
- Karmal (H)

87

Borassus flabellier L.

Hindi Name: Taad

English Name: Palm

Family : Arecaceae

Flowering: February - March

Fruiting April - August

Botanical Description: The plant is a large tree and grows upto 30-40m. in height. The trunk may have a circumference of 1.7 m at the base. The leaves are palmate, leathery, gray green and 1-3 m in wide with folded along at the midrib. Flowers are 90-150 cm long and branched. Fruits are ovoid, 15-20 cm in diameter with fibrous and fleshy mesocarp.

Ecology: The plant is frequently planted and naturalized in fields and waysides. The plant is native to tropical Africa but cultivated and naturalized throughout india.

Use: The fruits have food value and very much nutritive, rich in sugar and protein.

Nutritional Value: The fruits of the plant are rich in carbohydrate vitamins (C, B) and minerals (Iron, calcium). The nutritional value of the plant is presented below. 100g of the plant contains approximately:

Carbohydrate 10.9g

Protein 0.8g

Fat 0.1g

Vitamin C(5mg), B (0.02mg), B1(0.04mg), B3(0.3mg),

Mineral Ca(27mg), Fe(1mg), P(30mg),

Economical value: The stem is used as firewood and for construction purposes. Leaves are used to prepare hats, fan etc.

Borassus flabellier L.
Vernacular names :
- Tala (O)
- Palm (E)
- Taad (H)

88

Coccinia grandis (L.) Voigt

Hindi Name: Kundru

English Name: Lvy Gourd

Family : Cucurbitaceae

Flowering & Fruiting: Most part of the year

Botanical Description: The plant is a climbing herb with angular stem and long tuberous root. The leaves are classified as palmately simple with five lobes while the shape varies from the heart to pentagon form. The size of the leaves is approximately 5-10 cm in width and length. The tendrils are long, elastic with coil-like springy character that can wrap around the host to the entire length. The flower is large and white about 4 cm in diameter and contains five long tubular petals. The fruit is oblong, narrowed apically or at both ends. The raw fruit is green in color and turns bright red when ripened. The seeds of the plant are oblong and compressed.

Ecology: The plant is common in scrub forest, hedges and thickets, also cultivated. It is distributed throughout India.

Use: Ripe fruits are red in colour, eaten raw and cooked. Tender leaves are used as leafy vegetables by tribal people and also the tender fruits are eaten raw as it taste like cucumber.

Nutritional Value: The fruits of the plant are rich in carbohydrate, protein, vitamin B and minerals (calcium, potassium). The nutritional value of the plant is presented below. 100g of the plant contains:

Carbohydrate 3.1g

Protein 1.2g

Fat 0.1 g

Vitamin B1 (0.07mg), B2(0.08mg), B3(0.07mg) C(1.4mg)

Mineral Ca(40 mg), Fe(1.4mg), P(30mg)

Economic vlaue: Besides the fruit being used as a vegetable, it is eaten by birds and cattles.

Coccinia grandis (L.) Voigt
Vernacular names
- Kainchikakudi (O)
- Lvy gourd (E)
- Kundru (H)

89

Cucumis melo L.

Hindi Name: Kharbooza

English Name: Muskmelon

Family : Cucurbitaceae

Flowering: September-October

Fruiting: October-November

Botanical Description: The plant is an procumbent annual climber growing to 1.5m. The leaves are oval to kidney shaped with five to seven shallow lobes. Plants produce trailing vines and yellow flowers similar to cucumbers, but flowers may be male, female, or perfect. Fruit is spherical or ovoid, large, and can reach the size of a football, although it is generally more flattened. It is striped green or pale green in colour, ultimately turning yellow.

Ecology: It is frequently seen in waste ground, often cultivated. Many cultivated forms differing in size, skin and pulp colour and taste of the fruits are sometimes cultivated. The plant is distributed throughout India.

Use: Melon is a very popular summer fruit in northern parts of India

Nutritional Value: The fruits of the plant are rich in Vitamins and minerals. The nutritional value of the plant is presented below. 100g of the plant contains about:

Carbohydrate 8.6g

Protein 0.84g

Fat 0.19g

Vitamin C(36.7mg), B(0.026mg) B1(0.017mg), B3(0.734mg)

Mineral Ca(9mg), Mg(12mg), Fe(0.21mg)

Economic value: The plants are also eaten by cattles.

Cucumis melo L.
Vernacular names
- Phutikakaudi (O)
- Muskmelon (E)
- Kharbooza (H)

90

Dillenia indica L.

Hindi Name: Chalta
English Name: Elephant Apple

Family : Dilleniaceae

Flowering: May- June

Fruiting: September-February

Botanical Description: The plant is a moderate sized evergreen tree growing up to 15m tall. The leaves are deep green in colour, oblong to lanceolate and 15-36 cm in long. The flowers are white, large, up to 5 inches across. The fruit is rounded, 5-12 cm in diameter aggregate of 15 carpels with each carpel containing 5 seeds embedded in the pulp.

Ecology: It is frequently planted and often grown in gardens. The plant is widely distributed in evergreen forests throughout India.

Use: The ripe fruits are collected by local people and the fleshy sepals are sour in taste, used for making soup and pickles.

Nutritional Value: The fruits of the plant are rich in protein and fibre. The nutritional value of the plant is presented below. 100g of the plant contains:

Carbohydrate 3.1g

Protein 11g

Fat 0.12g

Fibre 1.6g

Vitamin C (4mg), B_1 (0.07mg)

Mineral Ca(10 mg), K (26mg), Fe (1.4 mg)

Economic value: The leaves and tree trunk is used as firewood. The fruits have a good market value.

Dillenia indica L.
Vernacular names
- Awoo (O)
- Elephant apple (E)
- Chalta (H)

91

Dioscorea alata L.

Hindi Name: Chupri Alu

English Name: Asiatic Yam

Family : Dioscoreaceae

Flowering & Fruiting: Throughout the year

Botanical description: The plant is a vigorous, perennial, climbing plant, producing annual stems up to 15 metres long from a very large tuberous rootstock. Tubers of the plant are one to several, polymorphic, variously branched or lobed and hairy with small rootlets. The stems scramble over the ground, or twine into the surrounding vegetation. The stem is quadrangular, winged, and twines right-handed. Leaves are opposite, 7-18 x 4-10 cm, broadly ovate, base cordate with acuminate apex. Flowers are laxly arranged, sessile with 1-1.5 mm long and ovate bracts and fleshy tepals.

Ecology: It is cultivated and also naturalized in almost all tropics and in subtropics and distributed throughout India. It is either planted or also found wild. The plant grows in the rainy season, loves to grow in humus rich soil.

Use: Root is usually used as vegetable after cooking/boiled. It is a very much nutritive food but it can be toxic if eaten raw.

Nutritional Value: The nutritional value of the plant is presented below. 100g of the plant (tuber) contains about:

Carbohydrate 27.8g

Protein 1.53g

Vitamin: A (138 IU) C(17.1mg), B_1(0.552mg), B_2(0.032mg), B_3(0.112mg), E(0.35mg)

Mineral: Ca(17mg), P(55mg), Fe(0.54mg), Mg(21mg)

Economic value: The plant has also medicinal value as it controls blood sugar and avoid diseases like constipation, leprosy and tumours. The root paste is applied over the body in fever. The leaves of the plant are eaten by cattles. Yams tips of the climbers are eaten as curry.

Dioscorea alata L.
Vernacular names
- Khamb alu (O)
- Asiatic yam/Water yam (E)
- Chupri alu (H)

92

Dioscorea bulbifera L.

Hindi Name: Zimikand

English Name: Air Yam

Family : Dioscoreaceae

Flowering & Fruiting: September-October

Botanical description: Aerial yam is a glabrous-leafed, non-spiny, perennial climbing plant producing annual stems up to 10 metres long from a woody, tuberous rootstock. These stems scramble over the ground, or twine into the surrounding vegetation. The stems twine left-handed and produce aerial axillary bulbils. The plant produces tubers; however the bulbils which grow at the base of its leaves are more important food products. They are about the size of potatoes (hence the name air potato) weighing from 0.5 to 2 kg. The flowers are whitish or pinkish in colour.

Ecology: The plant is found in wild and often cultivated in tropical regions. The plant loves to grow in wet season.

Use: Some varieties are eaten raw while some require soaking or boiling for detoxification before eating. It is not grown much commercially since the flavour of other yams is preferred by most people. However, it is popular in home vegetable gardens because it produces a crop after only 4 months of growth and continues for the life of the vine, as long as two years. It is a much nutritive food but it can be toxic if eaten raw.

Nutritional Value: The nutritional value of the plant is presented below. 100g of the plant (root) contains about:

Carbohydrate 32.5g

Protein 2.64g

Vitamin: A (112 IU) C(21.1mg), B_1(0.339mg), B_2(0.086mg), B_3(0.213mg), E(0.72mg)

Mineral: Ca(21mg), P(32mg), Fe(0.39), Mg(12mg)

Economic value: The tuber of the plant is used as medicine against diseases like dysentery and piles.

Dioscorea bulbifera L.
Vernacular names
- Pita kanda (O)
- Air yam (E)
- Zimikand (H)

93

Dioscorea pentaphylla L.

Hindi Name: Kanta Alu

English Name: Fire Leaf Yam

Family : Dioscoreaceae

Flowering & Fruiting: August-October

Botanical description: Stem twining to the left, terete, tomentose. Leaves 3-foliolate, palmately divided into 3-5 subequal leaflets, broadly ovate, oblique at base, abruptly acuminate; petiole 15-18 cm long. Spikes are 30 cm long, densely tomentose, distantly flowered fascicles at nodes on long peduncle. Female flowers sessile; bracts 4 mm long, ovate; tepals 2 x 1 mm, oblong, acute, hairy. Capsule 25 x 15 mm, oblong, tomentose. Male flowers not seen.

Ecology: The plant is usually found as wild in Semi-evergreen, moist and dry deciduous forests. The plant grows luxuriously during wet season in the humus rich soil.

Use: It is consumed by boiling or cooking

Nutritional Value: The nutritional value of the plant is presented below. 100g of the plant (root) contains about:

Carbohydrate 40.5g

Protein 3.12g

Vitamin C(18.1mg), B_1(0.292mg), B_2(0.056mg), B_3(0.302mg), E(0.64mg)

Mineral Ca(18mg), P(12mg), Fe(0.39), Mg(22mg)

Economic value: Tuber paste is applied on joint pain, rheumatism. Sometimes chips are also prepared from the root.

Dioscorea Pentaphylla L
Vernacular names
- Masia kanda (O)
- Five Leaf Yam (E)
- Kanta alu (H)

94

Dioscorea tomentosa L.

Hindi Name: Kanta Alu

English Name: White Yam

Family : Dioscoreaceae

Flowering & Fruiting: July-December

Botanical description: The plant is a perennial climbing plant. Tubers are irregular, usually long ovoid, transverse section white when fresh, becoming brown. Roots fibrous. Leaves are compound, palmately divided into 3-5 leaflets, and alternately arranged on the twining stems. Male flowers are nearly stalkless and Female spikes are simple or branched. Capsules are thinly leathery, sparsely hairy and turns black at maturity.

Ecology: The plant is widely cultivated in tropical Asia for its edible root. It is also harvested from the wild and usually found in forests. The plant loves to grow in wet season in humus soil.

Use: It is consumed after boiling or cooking. the tuber of the plant and is used as medicine for boils and swellings.

Nutritional Value: The nutritional value of the plant is presented below. 100g of the plant (root) contains about:

Carbohydrate 22.5g

Protein 2.12g

Vitamin C(10.1mg), B_1(0.102mg), B_2(0.018mg), B_3(0.112mg), E(0.26mg)

Mineral Ca(0.82mg), P(10mg), Fe(0.20),

Economic value: The roots are used for preparing chips during off season.

Dioscorea tomentosa L.
Vernacular names
- Taraga kanda (O)
- Kanta alu (H)
- White yam (F)

95

Diospyros malabarica (Desr.) Kostel

Hindi Name: Gaab

English Name: Gaub persimmon

Family : Ebenaceae

Flowering: March-April

Fruiting: March-April (Next Year)

Botanical Description: The plant is an evergreen tree with a spreading crown. It can grow up to 37 m tall, with a trunk girth of 2 m. The bark is black, smooth, and the inner bark turns bluish on exposure to sunlight. Leaves are oblong and glossy. The male flowers are formed in 3-5 flowered cymes in leaf axils. Female flowers are solitary. Fruits are round, up to 3.5 cm in diameter, yellow when ripe with a juicy pulp.

Ecology: The plant is commonly seen in lowland rainforests, typically along rivers and streams and often found in shady and wet sites near streams in the forest

Use: Ripe fruits are soft, sweet in taste and are widely eaten by tribal communities.

Nutritional Value: The fruits of the plant are rich in Vitamins and minerals. The nutritional value of the plant is presented below. 100g of the plant contains about:

Carbohydrate 3.6g

Protein 4.8g

Vitamin A (3.42mg) C(1.8g), B(0.176mg) Mineral Ca(4.9mg), Fe(0.33mg)

Economic value: The fruit and bark is beneficial in treatment of diarrhoea and dysentery. The fruit pulp is also used as gum for repairing and book binding. The wood of the plant is very strong and used for construction purposes.

Diospyros malabarica (Desr.) Kostel
Vernacular names
- Mankada Kendu (O)
- Gaub persimmon (E)
- Gaab (H)

96

Diospyros melanoxylon Roxb.

Hindi Name: Tendu

English Name: Coromandel Ebony Persimmon

Family : Ebenaceae

Flowering: April-May

Fruiting: (Next Year) March-April

Botanical Description: The plant is a medium-sized tree growing up to 25 m, height and 1.9 m girth. The bark is black in colour, peeling in rectangular scales, which is an identifying feature. The leaves are eliptic-oblong, opposite or alternate, leathery, up to 35 cm long, velvety on both sides when young, becoming smooth above when fully grown. Male flowers are mauve in colour, sessile or nearly sessile in short peduncles, mostly 3-flowered. Female flowers mauve, mostly extra-axillary or sometimes solitary, axillary, opposite to each other, larger than the male flowers. Fruits olive green in colour, ovoid or globose 3-4 cm across, 1-8 seeded berries. Seeds are compressed, oblong and shiny.

Ecology: It is fairly common in open and dry deciduous forests and distributed in peninsular India.

Use: Ripe fruits are soft, sweet in taste and are edible. The fruits are also sold in tribal and rural markets by the local people.

Nutritional Value: The fruits of the plant are rich in Vitamins and minerals. The nutritional value of the plant is presented below. 100g of the plant contains about:

Carbohydrate 2.6g

Protein 4.3g

Vitamin A (4.82mg) C(2.8g), B(0.026mg) Mineral Ca(7.5mg), Fe(0.78mg)

Economic value: The timber is very hard and used as firewood. The young leaves of the plant are used for *bidi* making.

Diospyros melanoxylon Roxb.

Vernacular names

- Kendu (O)
- Coromandel Ebony persimmon (E)
- Tendu (H)

97

Ficus benghalensis L.

Hindi Name: Barh

English Name: Banyan Tree

Family : Moraceae

Flowering: Round the year

Fruiting April - June

Botanical Description: The tree is very large, growing up to 30 m tall, with many aerial roots that develop in to new trunks, so that the tree goes on spreading laterally. Indefinitely a single tree can cover a very wide area. The leaves are leathery, entire, ovate or elliptic, which grow up to 20-30 cm long with prominent lateral veins. The fruits are 1 to 2 cm in diameter, rounded in shape, without stalks arranged in pairs at leaf axils and it changes to bright red colour when ripens. The plant flowers round the year and bears fruits during April-June.

Ecology: It is fairly common seen in wild and sometimes planted. The plant is widely distributed throughout the plains of India.

Use: Ripe fruits are eaten in food scarcity period. Tribal peoples use it as food regularly.

Nutritional Value: The fruits of the plant are rich in carbohydrate vitamins (E, C and B) and minerals (Iron, calcium, Magnesium). The nutritional value of the plant is presented below. 100g of the plant contains:

Carbohydrate 19.18g

Protein 0.75g

Fat 0.30g

Vitamin E (0.11mg), C(2mg), K (4.7μg), B (0.050mg), B_1(0.060mg), B_3(0.400mg)

Mineral Ca(35mg), Fe(0.37mg), M(17mg), Mn(0.128 mg)

Economic Importance: The plant wood is used for preparation of agricultural implements. The trees provide shade on the road sides and planted for religious purposes. Fruits are aslo eaten by birds. The plant has got religious importance.

Ficus benghalensis L.
Vernacular names
- Baro (O)
- Banyan tree (E)
- Barh (H)

98

Ficus recemosa L.

Hindi Name: Goolar

English Name: Cluster Fig Tree

Family : Moraceae

Flowering: March-June

Fruiting June-July

Botanical Description: It is a large sized tree with few short areal roots. Leaves are ovate or elliptic with 7-20×3.5-9 cm in size. The most distinctive aspect of this tree is the red, furry figs in short clusters, that grows directly out of the trunk of tree. The figs do carry hundred of flowers in it.

Ecology: The plant is fairly common in valleys and village surroundings. It is widely distributed throughout India.

Use: Ripe fruits are having a pleasant odour and eaten raw by local people.

Nutritional Value: The fruits of the plant are rich in carbohydrate, protein, vitamins (A, C) and minerals (Iron, calcium). The nutritional value of the plant is presented below. 100g of the plant contains:

Carbohydrate 15.84g

Protein 28.12g, Fat 1g

Vitamin A (200mg), C(5.3mg)

Mineral Ca(30.5mg), Fe(315mg), P(103mg), Na(329 mg)

Economic value: The plant also used for fire wood purpose. The fruits of he plant are used to prepare chariot by children in fun making.

Ficus recemosa L.
Vernacular name
- Dimiri (O)
- Cluster fig tree (E)
- Goolar (H)

99

Limonia acidissima Groff

Hindi Name: Kaitha

English Name: Wood Apple

Family : Rutaceae

Flowering: February-April

Fruiting: November-January

Botanical Description: The plant is an erect, slow-growing tree with a few upward-reaching branches. The bark is ridged, fissured and scaly with sharp spines. The leaves are alternate, 3 to 5 in long, dark-green in colour, leathery, often minutely toothed, blunt or notched at the apex, are dotted with oil glands and slightly lemon-scented when crushed. Flowers are greenish or reddish in colour, 1/2in across, are borne in small, loose, terminal or lateral panicles. The tree is mostly known for its hard woody fruit, size of a tennis ball, round to oval in shape. The pulp is brown, odorous, resinous, acid or sweetish, with numerous small, white seeds scattered through it.

Ecology: It is frequently found in open forest often planted and domesticated, usually seen near to villages and distributed throughout India.

Use: The pulp of the fruits is used to prepare chutney/pickles.

Nutritional Value: The fruits of the plant are rich in Vitamins and minerals. The fruit is a rich source of vitamin C. The nutritional value of the plant is presented below. 100g of the plant contains about:

Carbohydrate 3.1g

Protein 5.8g

VitaminA (0.04mg) C(180mg), B_2(0.23mg), B_1(0.31mg)

Mineral Ca(7.5mg), Fe(0.78mg)

Economic value: The plant has medicinal value as it having anti-microbial properties and also used for treating digestive problems. The pulp of the fruit, especially when unripe, is used in the treatment of diarrhoea and dysentery

Limonia acidissima Groff

Vernacular names

- Kaitha (O)
- Wood Apple (E)
- Kaith (H)

100

Madhuca longifolia (J.Koenig ex L.) J.F.Macbr.

Hindi Name: Mahua

English Name: Indian Butter Tree

Family : Sapotaceae

Flowering: February-April

Fruiting: May-July

Botanical Description: The plant is a large or moderate-sized tree, smooth and brown bark with pubescent or tomentose branchlets. The leaves are clustered near the ends of the branches, elliptic, oblong or ovate and acuminate. The flowers are cream coloured, fleshy, on rusty tomentose pedicel and clustered at the end of the usually leafless branches. The fruits are fleshy, green berries, quite large and containing from one to four shiny, brown seeds.

Ecology: It is common in margins of tropical and subtropical forests and also found in village sides and widely distributed throughout India.

Use: The flowers are edible, which is very delicious and nutritive.

Nutritional Value: The nutritional value of the plant is presented below. 100g of the plant(flowers)contains about:

Carbohydrate 2.3g

Protein 0.88g

Vitamin A (512 IU) C(172mg), B_2(0.23mg)

Mineral Ca(45mg), P(22mg), Fe(1.1mg)

Economic value: The bark is used medicinally for the treatment of leprosy. It is also used externally to treat itchy skin and bleeding gums. The oil from the seeds is used in the treatment of skin diseases. The fermented flowers are used in the preparation of local wine, the flowers are eaten by wild animals like boars and bears.

Madhuca longifolia (J.Koenig ex L.) J.F.Macbr.
Vernacular names
- Mahula (O)
- Indian Butter Tree (E)
- Mahua (H)

101

Manilkara hexandra (Roxb.) Dubard

Hindi Name: Khirni

English Name: Milk Tree

Family : Sapotaceae

Flowering: November-January

Fruiting: March-June

Botanical Description: The plant is a tree growing up to 3-12 m tall, with gray bark and smooth branchlets. The leaves are alternately arranged often closely clustered towards the end of branchlets, obovate to obovate-elliptic leaf blade with 0.8-2 cm leaf-stalks. Flowers are white or light yellow in colour, about 4 mm in size and arise in fascicles in leaf axils. Berry is oblong to ellipsoid, 1- 2 seeded and yellow when ripe.

Ecology: It is fairly abundant in open forests and also in uplands of mangrove forests and distributed throughout India.

Use: The ripe fruits are sweet in taste, eaten fresh or dried, which is very nutritive.

Nutritional Value: The nutritional value of the plant is presented below. 100g of the plant contains about:

Carbohydrate 3.6g

Protein 0.48g

Vitamin A (675IU) C(15.67mg) Mineral Ca(83mg), P(17mg), Fe(0.92)

Economic value: The leaves are eaten by cattle. Seeds are used for production of edible oil

Manilkara hexandra (Roxb.) Dubard
Vernacular names
- Khirakoli (O)
- Milk tree (E)
- Khirni (H)

102

Momordica dioica Roxb. ex Willd.

Hindi Name: Ban Karela

English Name: Spiny Gourd

Family : Cucurbitaceae

Flowering: August-September

Fruiting: September-November

Botanical Description: The plant is a large perennial climber which grows up to 3-5 m. The leaves are simple, alternately arranged, ovate and 3-5 lobed. The flowers are yellow, 4-6 cm across, solitary on axillary stalks. Fruit is an ellipsoid, 3-6 cm long, densely covered with soft spines.

Ecology: It is common in hedges and thickets, apparently wild and distributed throughout India. The plant grows in wet season. Though the plant is seen in the wild, it is being cultivated commercially because of market value.

Use: The unripe fruits are edible and used as a vegetable which is very nutritive.

Nutritional Value: The nutritional value of the plant is presented below. The fruit are rich in calcium and phosphorus. 100g of the plant contains about:

Carbohydrate 7.7g

Protein 3.1g

Fibre 3.0g

Vitamin A (0.624mg) C(16.20mg), B_1(1.3mg), B_2(0.56mg), B_3(0.86mg)

Mineral Ca(75mg), P(18.8mg), Fe(0.87)

Economic value: The root paste is applied over the body of patient suffering from fever.

Momordica dioica Roxb. ex Willd
Vernacular names
- Kankad (O)
- Spiny gourd (E)
- Ban karela (H)

103

Musa paradisiaca L.

Hindi Name: Kela

English Name: Banana

Family : Musaceae

Flowering & Fruiting: Most parts of the year

Botanical description: Banana is a tropical small size tree, with large leaves which overlapping bases from the trunk. The leaves are oblong and 1.2-1.5 m in length. The stem reaches a height of about 10-30 feet. Flower develops from the centre of the crown. Only female flowers develop into a banana fruit that varies in length from 4 to12 inches. Each banana plant bears fruits only once. The plant bears flowers and fruits in most parts of the year.

Ecology: The plant is a large herb, indigenous to Odisha, Bihar and eastern Himalayas. The plant is cultivated throughout India. The plant loves to grow in rainy season in humus rich soil. The plant is propagated through stems.

Use: The plant has lots of economic value. The fruit is eaten, both in raw and ripen form and is rich in potassium. The flower of the plant is also consumed by removing the spathe followed by cooking, which is a good source of vitamin E and other nutrients. The central stock is rich in iron and cooked in rural areas.

Nutritional Value: The nutritional value of the plant is described below. 100g of the plant (fruit) contains approximately:

Carbohydrate 22.84g

Protein 1.09g

Fat 0.33g

Vitamin B_1-0.31mg, B_2-0.073mg,

B_3-0.665mg, C-8.7mg, E-0.10mg

Mineral Ca-5mg, Fe-0.26mg, P-22mg, Mg-27mg, Mn-0.270mg,

Zn-0.15mg

Economic value: The leaves of the plant are used as plates, and sometimes as firewood when dried. The leaves has also got religious important and used in different festivalss & pujas. Traditionally the leaves stocks after burnt are used for washing clothes.

Musa paradisiaca L.
Vernacular names
- Kadali (O)
- Banana (E)
- Kela (H)

104

Nelumbo nucifera Gaertn.

Hindi Name: Kamal

English Name: Lotus

Family: Nelumbonaceae

Flowering & Fruiting: July-December

Botanical description: It is a perennial aquatic plant, growing from a tuberous rootstock lying in mud at the bottom of water bodies, lakes and ponds. The plant produces large leaves which float on the surface of the water. Leaves are large, simple, glabrous, solitary, beset with scattered hard minute papillae and rounded in shape and covered with a network of microscopic hairs. Flowers are large, pink or white in colour with expanding and emerging above water. Fruit is 1-2 cm across with flat-topped seed pod divided into compartments. Seeds globose, 1 mm across and brown in colour

Ecology: The plant is usually found in water tanks, ponds, ditches, puddles and large lakes. It is distributed throughout India. Plant grows luxuriously in wet season

Use: The root of the plant is consumed as raw or cooked. It is also a source of starch or arrowroot and contains about 1.7% protein, 0.1% fat, 9.7% carbohydrate. Seeds are also edible as raw or cooked. Young leaves and their petioles are used as leafy vegetables in some localities.

Economic Value: The flower of the plant is considered as sacred flower as it is the symbol of Hindu deity *Maa Laxmi* and also used in different pujas and festivals (specialy in Laxmipuja). Because of its holy importance, Lotus is considered as the national flower of India. The leaves are used as plates for eating purpose.

Nelumbo nucifera Gaertn.
Vernacular names
- Padma (O)
- Lotus (E)
- Kamal (H)

105

Nymphaea alba L.

Hindi Name: Kanval

English Name: Water Lily

Family : Nymphaeaceae

Flowering: Throughout year

Fruiting: Throughout year

Botanical description: It is an aquatic perennial herb, having erect perennial root stocks that anchor it to the mud in the bottom of the water bodies. Leaves vary from being elliptic-oval to almost circular 15-26 cm across, fine hairs on upper side and always remain at floating stage. Flowers are mostly white, occasionally pink or red reaching up to 15 cm in diameter.

Ecology: The plant is commonly found in lakes and ponds. It is distributed throughout the warmer parts of India.

Use: The root of the plant is consumed as raw or cooked. Often boiled and eaten as starchy vegetable during scarcity periods. Unripe fruits are also eaten. Seeds are fried in sand for getting Khaie/Liya.

The flowers have a good market value used for worshiping goddess Laxmi.

Nymphaea alba L.
Vernacular names
- Kain (O)
- Water lily (E)
- Kanval (H)

106

Phoenix sylvestris (L.) Roxb.

Hindi Name: Khajur
English Name: Date Palm

Family : Arecaceae

Flowering: February-March

Fruiting: April-May

Botanical Description: The plant is a moderate sized dioecious tree which grows up to 4-10 m. in height and 30 cm diameter with a large crown and rough trunk covered with persistent leaf bases. The leaves are 3 m. long, gently recurved on 1 m petioles with spines near the base. The leaf crown grows to 5 m wide and 7-10 m tall containing up to 100 leaves. The inflorescence grows upto 1 m with white flowers. The single seeded fruit ripens to orange-yellow.

Ecology: It is commonly, wild or cultivated throughout India. It thrives from the plains to the coasts in low-laying waste lands, scrub forest and water logged areas. The plant loves to grow in dry and desert areas.

Use: The fruits of the plant are sweet and edible. It is very much nutritive and can be used to make juices, jellies, jams etc. and eaten as berries.

Nutritional Value: The fruit is rich in carbohydrate, vitamins and minerals. The nutritional value of the plant is described below. 100g of the plant (fruit) contains approximately:

Carbohydrate 74.97g

Protein 1.81g

Fat 0.15g

Vitamin A-149 IU,B1-0.050mg, B2-0.060mg, B3 1.610mg, K-2.7μG

Mineral Ca-64mg, Fe-0.90mg, P-62mg, Mg-54mg, Zn-0.44mg

Economic Value: In coastal areas, the tree trunk is used for making houses. The leaves are used for making thatching and making mats, hats, toys etc.

Phoenix sylvestris (L.) Roxb.
Vernacular name
- Khajuri (O)
- Date palm (E)
- Khajur (H)

107

Phyllanthus acidus (L.) Skeels

Hindi Name: Harfarauri

English Name: Star Gooseberry

Family : Phyllanthaceae

Flowering: May-June

Fruiting: June-September

Botanical Description: The plant is a small deciduous tree reaching about 25-30 ft in height. Leaves are compound, 14-25 inches long, crowded at the ends of the branches leaflets 2-3.5 inches long, alternately arranged along the rachis, ovate or obliquely ovate, acute with rounded or wedge-shaped base. Flowers are very minute, in short dense spike-like clusters arising from nodules along the branches. Fruit is pendulous, in small clusters from the branches, round or slightly flattened at the poles, with shallow or deep ribs (usually 5) 0.75 inch across.

Ecology: It is commonly found in coastal forests, open places and often grown in gardens.

Use: The fruits are consumed as raw or cooked. The mature fruits are also eaten fresh but usually they are sprinkled with salt to neutralize the acidity. It is rich in vitamin C with a very acid flavour, it is usually more used in relishes or chutneys, or cooked as a flavouring in desserts. The fruit is sometimes used as a substitute for tamarind. The fruit is yellow when fully ripe but turns red upon cooking.

Nutritional Value: The fruit is rich in vitamins and minerals. The nutritional value of the plant is described below. 100g of the plant (fruit) contains approximately:

Carbohydrate 0.64g

Protein 0.115g

Fat 0.52g

Vitamin B1-0.01mg, B2-0.05mg, C-8mg

Mineral Ca-5mg, Fe-0.4mg, P-23g,

Economic Value: The juice of the fruit is instilled in the eyes to treat eye-troubles. An extract from the root is used to cure skin diseases especially to bring relief from itching.

Phyllanthus acidus (L.) Skeels
Vernacular name
- Nara Koli (O)
- Star Gooseberry (E)
- Harfarauri (H)

108

Phyllanthus emblica L.

Hindi Name: Amla

English Name: Indian Gooseberry

Family : Phyllanthaceae

Flowering: March-May

Fruiting: November-February

Botanical Description: The plant is a small to medium sized deciduous tree, reaching 8 to 18 m in height. It is known for its edible fruits of the same name. The tree has crooked trunk and spreading branches. The leaves are simple, nearly stalkless and closely set along slender branchlets. The leaves are often mistaken for leaflets of pinnate leaves. The flowers are small, greenish-yellow or pinkish, having six segments, but no real petals. The fruit is nearly spherical, light greenish yellow, quite smooth and hard on appearance, with 6 vertical stripes or furrows

Ecology: The plant is commonly found in mixed forests, dry forests, dry open sparse forests or scrub and village groves.

Use: The fruits are consumed as raw or cooked. An acid, rather astringent flavour, they are not often eaten raw unless accompanied with sugar, salt or chillies to moderate those. The astringency can be removed by steeping the fruits in brine for a few days. The fruits are more commonly used to make jams, pickles, jellies, tarts, chutneys etc.

Nutritional Value: The fruit is a good source of Vitamin C. The nutritional value of the plant is described below. 100g of the plant (fruit) contains approximately:

Carbohydrate 10.18g

Protein 0.88g

Fat 0.58g

Vitamin B1-0.140mg, B2-0.030mg, B3-0.300mg, C-27.7mg

Mineral Ca-25mg, Fe-0.31mg, P-27mg, Mg-10mg, Mn-0.144mg

Economic value: The fruit has got medicinal value and are traditionally used as an expectorant, antipyretic, diuretic, antidiarrhoeal etc. in stomach and digestive disorders. The fruit powders are used in Ayurvedic medicine preparation. The dry fruits are used for pickle preparation.

Phyllanthus emblica L.
Vernacular name
- Aanla (O)
- Indian Gooseberry (E)
- Amla (H)

109

Pithecellobium dulce (Roxb.) Benth.

Hindi Name: Jangal Jalebi

English Name: Sweet Tamarind

Family : Mimosaceae

Flowering: January-March

Fruiting: April-June

Botanical Description: The plant is a large, nearly evergreen perennial tree that grows up to 20 m or more in height. It has a broad crown, up to 30 m across, and a short trunk, up to 1 m thick. Leaves are deciduous but foliage is persistent, as the new leaves appear while the old ones are being shed, so that the tree looks like an evergreen. Flowers yellowish-white in colour and are borne in small spherical glomerules of about 1 cm in diameter, forming short axillary panicles of 5-30 cm in length. Pods are greenish-brown to red or pinkish, rather thin with 10-15 cm long with irregular shape and flattened. There are about 10 seeds per pod, shining black in a white pulpy aril.

Ecology: It is commonly growing in dry, brushy or thinly forest plains or hillsides, wastelands and often in coastal thickets.

Use: The pods with a sweet, pulpy flesh that surrounds the seed, are edible as raw or by cooking. It is also used as fodder for goats, sheeps and cattles.

Nutritional Value: The fruit is rich in carbohydrate, vitamins and minerals. The nutritional value of the plant is described below. 100g of the plant(fruit) contains approximately:

Carbohydrate 7.6g Protein 1.3g Fat 0.2g

Vitamin A-5µg B_1-0.10mg, B_2-0.4mg, B_3-0.2mg, K-2.7µg, C-56mg

Mineral Ca-5mg, Fe-0.2mg, P-18mg, Mg-8mg, Zn-0.44mg

Economic value: The bark of the root is used as a good remedy for diarrhoea and dysentery. The leaves together with salt do cure indigestion.

Pithecellobium dulce (Roxb.) Benth.
Vernacular name
- Sima kaian (O)
- Manilla Tamarind/Sweet Tamarind (E)
- Jangal Jalebi (H)

110

Syzygium cumini (L.) Skeets

Hindi Name: Jamun

English Name: Java Plum

Family : Myrtaceae

Flowering: April-May

Fruiting: July-August

Botanical description: The plant is an evergreen tropical tree, 50 to 100 ft. tall, with oblong opposite leaves that are smooth, glossy and having a terpentine smell. It has fragrant white flowers in branched clusters at stem tips and purplish-black oval edible berries. The leaves are antibacterial, and are used for strengthning the teeth and gums. The fruit is ellipsoid or oblong, variable in size and shape, usually up to 2.5 cm long, black, shining and juicy when fully ripe. Seeds always one, with thick rounded fleshy cotyledons, closely appressed on their inner faces.

Ecology: It is fairly common in tropical and dry deciduous forests, also grown in roadsides and villages and distributed throughout India.

Use: Seeds are sweet, sub-acidic, tonic, usually and are used as medicine for diabetes, diarrhoea and ringworm infection. The bark is astringent, sweet sour, diuretic and digestive in nature. Fruits are used to prepare squashes, jellies and jams.

Nutritional Value: The nutritional value of the plant is presented below. The fruits are having good antioxidant value. 100g of the plant contains about:

Carbohydrate 9.7g

Protein 6.5g

Vitamin: C(49mg), B_1(2.6mg), B_2(1.56mg)

Mineral: Ca(31mg), P(1.72mg), Fe(4.66gm)

Economic value: Both the seeds and the fruits are diuretic and have important in curing diarrhea and dysentery. The seeds also reduce blood sugar levels and are useful in the treatment of diabetes. The trunk is used as firewood especially in brick kylns. The leaves are eaten by goat. Seeds have ayurvedic medicinal value.

Syzygium cumini (L.) Skeets
Vernacular names
- Jamu (O)
- Java plum (E)
- Jamun (H)

111

Trapa natans var. bispinosa (Roxb.) Makino

Hindi Name: Paniphal

English Name: Water Chesnut

Family : Lythraceae

Flowering: August-September

Fruiting: November-December

Botanical description: The water chestnut is an annual aquatic herb with a rosette of floating leaves and submerged, anchored into the mud by very fine roots. It has two types of leaves, finely divided, feather-like submerged leaves borne along the length of the stem, and undivided floating leaves borne in a rosette at the water's surface. The floating leaves have saw-tooth edges and are ovoid or triangular in shape, 2-3 cm long, on inflated leaf-stalks 5-9 cm long, which provide added buoyancy for the leafy portion. Flowers are white in colour, with Four petals and are borne for a long period. The fruit is a nut with barbed spines. Seeds can remain viable up to 12 years, although most germinate within the first two years.

Ecology: The plant is frequently in slow-moving rivers, tanks, lakes, swamps and ponds forming dense cover on water surface.

Use: The nut tastes sweet and delicious. It is eaten raw or after cooking and roasting. The nuts are nutritious and good source of minerals.

Nutritional Value: The plant is rich in vitamin B complex and protein, besides iron, calcium and phosphrous. The nutritional value of the plant (fruit) is described below. 100g of the plant contains approximately:

Carbohydrate 7.1g
Protein 0.10g
Fat 0.06g

Vitamin B_1-0.27mg, B_2-0.28mg, B_3-0.58mg, B_6-0.27mg

Mineral Ca-102mg, Fe-3083mg, P-325mg

Economic Value: The fruit of the plant have a good market value because of its nutritional importance.

Trapa natans var. *bispinosa* (Roxb.) Makino
Vernacular names
- Pani singada (O)
- Water Chestnut (E)
- Paniphal (H)

112

Ziziphus oenopolia (L.) Mill.

Hindi Name: Makai

English Name: Jackal Jujube

Family : Rhamnaceae

Flowering: July-September

Fruiting: October-January

Botanical Description: Jackal Jujube is a very thorny straggling shrub with rusty-velvety young branches with paired thorns. Thorns are, one straight and the other recurved. Alternately arranged simple leaves are ovate to ovate-lancelike, often oblique, with three prominent nerves and numerous transverse nervules. Tiny green flowers are borne in nearly stalkless velvety cymes in leaf axils. Fruits are spherical or obovoid drupes, black, shining, seeds woody.

Ecology: The plant is usually found in subtropical forests, thickets, wastelands, hedges etc.

Use: The fruits are edible. It tastes sweet to acidic when ripes, usually eaten as berries and also consumed during scarcity period. They can be dried for off-season consumption.

Nutritional Value: The nutritional value of the plant is described below. The fruits are rich source of viamin B complex, calcium, magnesium, manganese and zinc. 100g of the plant (fruit) contains approximately:

Carbohydrate 20.53g

Protein 1.2g

Fat 0.20g

Vitamin B1-0.020mg, B2-0.040mg, B3-0.900mg, C-69mg,

Mineral Ca-21mg, Fe-0.48mg, P-23mg, Mg-10mg, Mn-0.084mg, Zn-0.05mg

Economic value: Often pickles are prepared from ripened seeds. The plants are planted along the hedges to protect the crop(s).

Ziziphus oenopolia
Vernacular name
- Kantei koli (O)
- Jackal Jujube (E)
- Makai (H)

List of Wild Food Plants with their Part(s) and Mode of Consumption

Table 1: List of Wild Food Plants with their part(s) used and Mode of Consumption

Sl.No	Name of the Plant	Family	Common Name (English)	Local (Odia) Name	Habit	Part Used	Mode of Consumption
1.	*Achyranthes aspera* L.	Amaranthaceae	Prickly Chaff Flower	Apamaranga	H	L	Young leaves and shoots are collected, roasted and eaten
2.	*Acalypha indica L.*	Euphorbiaceae	Indian Copperleaf	Mukta jhuri	H	L	Leaves are fried with oil, then eaten
3.	*Aerva lantana* (L.) Juss	Amaranthaceae	Mountain Knot Grass	Paunsia	H	L	Leaves are collected, fried and eaten
4.	*Allium cepa* L.	Amaryllidaceae	Onion	Piaja	H	L	Young leaves and shoots are collected, fried then eaten
5.	*Allium sativum* L.	Amaryllidaceae	Garlic	Rasuna	H	L	Young leaves and shoots are collected, fried then eaten
6.	*Allmania nodiflora* R.Br.	Amaranthaceae	Node Flower Allmania	Chadheimundia saga	H	L	Leaves and young shoot are roasted with mustard oil and then eaten.
7.	*Alternanthera ficoidea* (L.) Sm.	Amaranthaceae	Sanguinarea	Bana madaranga	H	L	Young leaves and shoots are collected, fried then eaten
8.	*Alternanthera sessilis* L.R. Br.	Amaranthaceae	Sessile Joyweed	Madaranga	H	L	Leaves and young shoot are roasted with mustard oil and then eaten
9.	*Amaranthus blitum* L.	Amaranthaceae	Purple amaranth	Kosila	H	L	Leaves and young shoots are cut into small pieces, cooked with salt and chilly and then eaten
10.	*Amaranthus caudatus* L.	Amaranthaceae	Raddish Amaranath	Khada saga	H	L	Leaves and young shoots are cut into small pieces, roasted then eaten
11.	*Amaranthus spinosus* L.	Amaranthaceae	Prickly amaranth	Kanta leutia	H	L	Leaves and young shoots are cut into small pieces, cooked with salt and chilly and then eaten.
12.	*Amaranthus tricolor* L.	Amaranthaceae	Elebhrant-Head Amarnath	Lal Khada	H	L	Leaves are collected, fried/roasted then eaten

Contd.

13.	*Amaranthus viridis* L.	Amaranthaceae	Wild amaranth	Leutia	H	L	Leaves and young shoots are collected, fried then eaten
14.	*Andrographis paniculata* (Burm.f.) Nees	Acanthaceae	Creat	Bhuin nimba/Chireita	H	L	Leaves are collected, fried and eaten
15.	*Arisaema tortuosum (*Wall.) Schott	Araceae	Whipcord Cobra Lily	Bana nada	H	L	Young shoot/Fruit stalk is fried to prepare curry and eaten
16.	*Azadirachta indica* A.Juss	Meliaceae	Neem	Nimba	T	L	Fresh tender leaf along with flower is fried with mustard oil.
17.	*Bambusa bambos* (L.) Voss	Poaceae	Indian Throny Bamboo	Kanta baunsa	T	L	Young shoots (Karada) are cut into small pieces, cooked with salt and chilly and then eaten.
18.	*Basella alba* L.	Basillaceae	Indian spinach	Poi	C	L	Stem and leaves are used to prepare curry
19.	*Bauhinia purpurea* L.	Caesalpiniaceae	Geranium tree	Kuilari	T	L	Young shoots along with leaves are collected, cooked as curry or fried and taken.
20.	*Bauhinia vahlii* Wight & Arn.	Caesalpiniaceae	Camel's foot climber	Siali	C	L, Se	Young leaves are collected, cooked as curry and taken. Seeds are eaten after boiling or cooked as vegetable
21.	*Bauhinia variegata* L.	Caesalpiniaceae	Kachnar	Kanchan	T	L, Fl	Leaves are collected, cooked as curry and taken and flowers coated with rice powder are very good in taste after fried
22.	*Benincasa hispida* (Thunb.) Cogn.	Cucurbitaceae	White gourd	Panikakharu	C	L, Fr	Young shoots along with leaves are collected, cooked as curry or fried and then eaten. Fruit is commonly eaten as vegetable
23.	*Boerhavia diffusa* L.	Nyctaginaceae	Hog weed	Puruni saga	H	L	Tender leaves and young shoots are collected, fried/ roasted then eaten

Contd.

24.	*Brassica napus* L.	Brassicaseae	Rape	Sorisa saga	H	L, Se	Chopped leaves and stems are dried in shade, stored in an eaten. Seed is used as spices and extraction of edible oil
25.	*Brassica oleracea* L. var *botrytis* L.	Brassicaceae	Cauliflower	Phula kobi	H	L, Fl	Leaves are fried or made into curry and consumed. Inflorescence commonly eaten as vegetable
26.	*Brassica oleracea* L. var capitata L.	Brassicaceae	Cabbage	Banda/patra kobi	H	L	Leaves are fried separately or with other vegetables, then eaten
27.	*Brassica oleracea L. var. gongylodes L.*	Brassicaceae	Knol-Khol	Gantthi/Ol kobi	H	L	Leaves are fried separately or with other vegetables, then eaten
28.	*Celosia argentea* L.	Amaranthaceae	Quail grass	Ghurudi sag	H	L	Young leaves and shoots are collected, roasted then eaten
29.	*Centella asiatica* L.	Apiaceae	Indian pennywort	Thalkudi	H	L	Leaves and young shoots are collected, roasted then eaten
30.	*Chenopodium album* L.	Chenopodiaceae	White goosefoot	Bathua saga	H	L	Leaves and young shoots are cooked along with other vegetable adding salt and chilly to it.
31.	*Cicer arietinum* L.	Fabaceae	Chickpea	Buta saga	H	L, Se	Young leaves and shoots are collected, roasted then eaten. seeds consumed as raw or cooked
32.	*Cinnamomum tamala* (Buch.-Ham.) T.Nees & Eberm.	Lauraceae	Indian cassia	Teja patra	T	L	Leaves are used as spice in making curry.
33.	*Cleome gynandra* L.	Cleomaceae	Spider flower	Araka saga	H	L	Leaves and young shoots are collected, roasted then eaten
34.	*Cleome viscosa* L.	Cleomaceae	Asian spider flower	Bana sorisa	H	L	Leaves and young shoots are collected, fried/roasted then eaten
35.	*Colocasia esculenta* (L.) Schott	Araceae	Elephant ear taro	Saro	H	L, Tu	Leaves are collected, fried with oil then eaten. Tuber is consumed after cooking in curries

Contd.

36.	*Commelina attenuata* K.D.Koenig ex Vahl	Commelinaceae	Asian spider flower	Chhena saga	H	L	Leaves and young shoots are collected, fried/ roasted then eaten
37.	*Commelina benghalensis* L.	Commelinaceae	Day flower	Kansiri	H	L	Leaves and young shoots are collected, fried/ roasted then eaten
38.	*Corchorus capsularis* L.	Tiliaceae	White jute	Nalita	H	L	Tender leaves are collected, curry is prepared and eaten.
39.	*Coriandrum sativum* L	Apiaceae	Coriander	Dhania	H	L	Leaves and shoots are collected, added to curry for aroma and also chutney is prepared from the fresh leaves.
40.	*Cucurbita maxima* Duchesne.	Cucurbitaceae	Sweet gourd	Kakharu saga	C	L, Fr	Leaves and young stems are collected, roasted then eaten. Fruit is commonly consumed as vegetable
41.	*Dendrocalamus strictus* (Roxb.) Nees	Gramineae (Poaceae)	Bamboo	Baunsa	T	Sh	Young shoots are cut into small pieces, cooked with salt and chilly and then eaten
42.	*Emilia sonchifolia* (L.) DC. ex DC.	Asteraceae	Purple Sow Thistle	Musakani sag	H	L	Tender leaves and young shoots are collected, cooked then eaten
43.	*Enydra fluctuans* DC.	Asteraceae	Water cress	Harakuch	H	L	Leaves are cooked with mustard oil, chilly, salt and eaten
44.	*Eryngium foetidum* L.	Apiaceae	Long Coriander	Jangali dhania	H	L	Leaves and shoots are collected, added to curry for aroma and also chutney is prepared
45.	*Furcraea foetida* (L.) Haw.	Asparagaceae	Giant Cabuya	Muruga	S	L	Young shoots are collected, fried, then eaten.
46.	Glinus oppositifolius (L.)A.DC.	Molluginaceae	Gima	Pita saga	H	L	Young leaves and shoots are col lected, roasted with other veg etables then eaten

Contd.

47.	(L.)A.DC.*Hibiscus sabdariffa* L.	Malvaceae	Roselle	Kanuria anga	H	L	The leaves are very sour in taste. So, leaves are collected, cooked with water and othervegetables to prepare sour water 'sambar'
48.	*Hybanthus enneaspermus* (L.) F.Muell.	Violaceae	Spade Flower	Luni saga	H	L	Leaves and young shoots are collected, fried/roasted then eaten
49.	*Hydrolea zeylanica* (L.) Vahl	Hydroleaceae	Ceylon Hydrolea	Langulia	H	L	Leaves and young shoots are roasted and taken as food
50.	*Hygrophila auriculata* (Schumach.) Heine	Acanthaceae	Marsh Barbel	Khikri sag	H	L	Leaves are roasted with oil and then eaten
51.	*Ipomoea aquatica* Forssk.	Convolvulaceae	Swampy Morning Glory	Kalama saga	C	L	Leaves and tender shoots are col lected, cooked and then eaten
52.	*Lagenaria siceraria* (Molina)	Cucurbitaceae	Bottle gourd	Laoo	C	L, Fr	Leaves and young stems are cooked with other vegetables. Fruit is con sumed as vegetable
53.	*Lepidium sativum* L.	Brassicaceae	Garden cress	Himba saga	H	L	Leaves and young shoots are col lected, roasted then eaten
54.	*Leucas aspera* (Willd.) Link	Lamiaceae	Common Leucas	Gayisa	H	L	Leaves are eaten after frying or roasting
55.	*Limnophila indica* (L.) Druce	Plantaginaceae	Indian Marshweed	Keralata	H	L	Leaves and young shoots are roasted and taken as food
56.	*Marsilea minuta* L.	Marsileaceae	Dwarf Water Clover	Sunsunia	H	L	Tender leaves and young shoots are roasted and taken as food
57.	*Mentha spicata* L.	Lamiaceae	Spear mint	Podina	H	L	Young leaves paste adding salt chilly is used as chantey
58.	*Mitracarpus hirtus* (L.) DC.	Rubiaceae	Spear tropical girdlepod	Ganthia sag	H	L	Leaves are boiled with water and then eaten
59.	*Momordica charantia* L.	Cucurbitaceae	Bitter gourd	Kalara	H	L	Leaves are eaten after frying or roasting

Contd.

60.	*Moringa oleifera* Lam.	Moraginaceae	Drum stick	Sajana	T	L, Fr	Leaves are eaten after frying and also added to curries and fruit is also used to prepare curries, sambar etc.
61.	*Murraya koenigii* L. Spreng	Rutaceae	Curry tree leaf	Bhursanga	T	L	Leaves are used as spice in curry and give it a specific scent. Also young leaves used to prepare chutny.
62.	*Oroxylum indicum* (L.) Kurz	Bignoniaceae	Midnight horror	Phapni sag	T	L	Leaves are eaten after frying or roasting
63.	*Oxalis corniculata* L.	Oxalidaceae	Indian sorrel	Amliti saga	H	L	Leaves are cooked with mustard oil, chilly, salt with potato as deli cious curry to eat
64.	*Paederia foetida* L.	Rubiaceae	Skunk Vine	Pasaruni	H	L	Leaves are eaten after frying or roasting
65.	*Phyla nodiflora* (L.) Greene	Verbenaceae	Frog Fruit	Langli saga	H	L	Leaves are collected and taken after cooking
66.	*Pimpinella heyneana* (Wall. ex DC.) Kurz	Fabaceae	Heyne burnet	Charmali	H	L	Leaves along with flowers and seeds are collected, fried then eaten. The plant is having a special fragrance.
67.	*Pisum sativum* L.	Fabaceae	Pea	Matar saga	H	L	Leaves and young shoots are collected, fried with oil and other vegetables, eaten
68.	*Plectranthus mollis* (Aiton) Spreng.	Lamiaceae	Soft-Stem Mintleaf	Sakra/Gondri	H	L, Se	Leaves are fired with oil and chilly then eaten, also seeds are used to prepared sweets.
69.	*Polygonum plebeium* R.Br.	Polygonaceae	Alpine knot weed	Muthi saga	H	L	Leaves and young shoot are roasted with mustard oil and then eaten
70.	*Portulaca oleracea* L.	Portulacaceae	Common purslane	Balubalua saga	H	L	Tender leaves and shoots are collected, roasted then eaten

Contd.

71.	*Raphanus sativus* L.	Brassicaceae	Radish	Mulasaga	H	L, Rt	Leaves are fried along with other vegetables then eaten and modified root is used as vegetable and in salad
72.	*Rivea hypocrateriformis* Choisy	Convolvulaceae	Midnapore Creeper	Kindu sag	C	L	Leaves are eaten after frying or roasting
73.	*Rungia pectinata* (L.) Nees	Acanthaceae	Comb Rungia	Hasa ada/ Mati sag	H	L	Leaves are collected, fried with oil then eaten
74.	*Senna tora* (L) Roxb	Caesalpiniaceae	Sickle Senna	Chakunda Sag	H	L	Tender leaves and shoots are cooked and then taken as food
75.	*Solanum americanum* Mill.	Solanaceae	Black nightshade	Nunununia	H	L	Leaves are collected fried, then eaten
76.	*Solanum tuberosum* L.	Solanaceae	Patato	Allu saga	H	L,Tu	Fresh leaves are collected, roasted with oil and then eaten and Tubers are boiled or fried and eaten
77.	*Spinacia oleracea* L.	Chenopodiaceae	Spinach	Palanga	H	L	Leaves are fried with vegetables and also cooked with dal and potato to make dalma curry.
78.	*Talinum portulacifolium* (Forssk.) Asch. ex Schweinf.	Talinaceae	Ceylon Spinach	Matpoi	H	L	Leaves are collected fried with other vegetable then eaten
79.	*Trachyspermum ammi* (L.)	Apiaceae	Indian dill	Juani	H	L	Leaves and young shoots are collected, roasted then eaten. It is highly scented
80.	*Trianthema portulacastrum* L	Aizoaceae	Horse-Purslane	Puruni saga	H	L	Leaves and young shoots are collected, fried then eaten
81.	*Trigonella corniculata* L.	Fabaceae	Wild trifoli	Firingi saga/Kasturimethi	H	L	Leaves are fried along with other vegetables
82.	*Trigonella foenum-graecum* L.	Fabaceae	Fenugreek	Methi	H	L	Fresh leaves are collected and chutney is prepared or roasted and taken

Contd.

83.	*Vicia sativa* L.	Fabaceae	Black-Pod vetch	Anisaga	H	L	Leaves are roasted with oil and then eaten
84.	*Aegle marmelos* (L.) Correa	Rutaceae	Wood apple	Bela	T	Fr	Ripe fruits is eaten raw or by cooking and often used to prepare juice/drink
85.	*Artocarpus heterophyllus* Lam.	Moraceae	Jack fruit	Panasa	T	Fr	Ripe fruits are sweet, eaten raw and the unripe fruits are used as vegetables
86.	*Averrhoa carambola* L.	Oxalidaceae	Carambola	Karamonga	T	Fr	Eaten raw or cooked
87.	*Borassus flabellifer* L.	Arecaceae	Palm	Tala	T	Fr	Fruits are having sweet, jelly sap, consumedas raw/drink
88.	*Coccinia grandis (L.) Voigt*	Cucurbitaceae	Lvy gourd	Kainchikakudi	C	Fr	Fruits are consumes raw or cooked
89.	*Cucumis melo* L.	Cucurbitaceae	Muskmelon	Phutikakudi			very popular summer fruit in northern partsof India
90.	*Dillenia indica* L.	Dilleniaceae	Elephant apple	Oau	CT	FrFr	Ripe fruits are sour in taste, used for flavouring curries and to prepare differentsour dishes (pickles/jellies).
91.	*Dioscorea alata* L.	Dioscoreaceae	Asiatic yam	Khamb alu	C	Tu	Tuber Consumed after boiling or cooked with other vegetables
92.	*Dioscorea bulbifera* L.	Dioscoreaceae	Air yam	Pita kanda	C	Tu	Tubers are kept in water for a long time then boiled/cooked and eaten
93.	*Dioscorea pentaphylla* L.	Dioscoreaceae	Five Leaf Yam	Masia kanda	C	Tu	Tubers are eaten by cooking as vegetable
94.	*Dioscorea tomentosa* L.	Dioscoreaceae	White yam	Taraga kanda	C	Tu	Tubers are boiled, then cooked and eaten
95.	*Diospyros malabarica* (Desr.) Kostel	Ebenaceae	Gaub persimmon	Mankada kendu	T	Fr	Ripe fruits are soft, sweet in taste and eatenas raw
96.	*Diospyros melanoxylon* Roxb.	Ebenaceae	Coromandel Ebony persimmon	Kendu	T	Fr	Ripe fruits are sweet in taste and eaten as raw

Contd.

97.	*Ficus benghalensis* L.	Moraceae	Banyan tree	Bara	T	Fr	Ripe fruits are eaten in food scarcity period.
98.	*Ficus recemosa* L.	Moraceae	Clister fig tree	Goolar	T	Fr	Tribal peoples use it as food regularly
99.	*Limonia acidissima* Groff	Rutaceae	Wood Apple	Kaintha	T	Fr	The pulp of ripe fruits are sour in taste and used to prepare chutney/ jellies and often eaten as raw
100.	*Madhuca longifolia* (J.Koenig ex L.) J.F.Macbr.	Sapotaceae	Indian Butter Tree	Mahula	T	Fl	Flowers are edible, which is very delicious and nutritive and also used in the preparation of local wine
101.	*Manilkara hexandra* (Roxb.) Dubard	Sapotaceae	Milk tree	Khirakoli	T	Fr	Ripe fruits are eaten fresh as berries
102.	*Momordica dioica* Roxb. ex Willd.	Cucurbitaceae	Spiny gourd	Kankada	C	Fr	Fruits are consumed as vegetable by frying with oil and other vegetables
103.	*Musa paradisiaca* L.	Musaceae	Banana	Kela	T	Fr,	Fruit is eaten, both in raw and ripen form.
						Fl,	The flower of the plant is also consumed by
						St,	removing the spathe followed by cooking, and the central stock is rich in iron and cooked in rural areas.
104.	*Nelumbo nucifera* Gaertn.	Nelumbonaceae	Lotus	Padma	H	Rt,	Root is collected, cut into small pieces then
						Se	fried and eating. Seeds are eaten as raw or cooked
105.	*Nymphaea alba* L.	Nymphaeaceae	Water lily	Kain	H	St	Stem is collected, fried then eaten

Contd.

106.	*Phoenix sylvestris* (L.) Roxb.	Arecaceae	Date palm	Khajuri	T	Fr	Ripe fruits are sweet, eaten raw and can be preserved to prepare jellies/jams etc.
107.	*Phyllanthus acidus (*L.) Skeels	Phyllanthaceae	Star Gooseberry	Nara koli	T	Fr	Fruits are consumed as raw or cooked. The mature fruits are also eaten fresh with salt to neutralize \ the acidity
108.	*Phyllanthus emblica* L.	Phyllanthaceae	Indian Gooseberry	Amla	T	Fr	Fruit is eaten raw as berries and used to prepare pickles/jellies
109.	*Pithecellobium dulce* (Roxb.) Benth.	Mimosaceae	Sweet Tamarind	Sima Kaian	T	Fr	Fruit is a pod with sweet, pulpy flesh that surrounds the seed, are edible as raw or by cooking
110.	*Syzygium cumini* (L.) Skeets	Myrtaceae	Java plum	Jamu	T	Fr	Ripe fruits are widely eaten as berries
111.	*Trapa natans* var. *bispinosa* (Roxb.) Makino	Lythraceae	Water Chestnut	Pani singada	H	Fr	The nuts are eaten raw when fresh of consumed after cooking or boiling.
112.	*Ziziphus oenopolia* (L.) Mill.	Rhamnaceae	Jackal jujube	Kanteikoli	S	Fr	Ripe fruits are eaten raw mostly by the children as it is sweet and acidic in taste.

T: Tree, S: Shrub, H: Herb, C: Climber, L: Leaf, Fl-: Flower, Fr: Fruit, Sh: Shoot, Tu: Tuber, Rt: Root, Se: Seed

Index

E

F

G

H

I

J

K

L

M

N

O

P

R

W

Y

Z